A GUIDE

TO THE

MICROSCOPICAL EXAMINATION

OF

DRINKING WATER

A GUIDE

TO THE

MICROSCOPICAL EXAMINATION

OF

DRINKING WATER

BY

J. D. MACDONALD, M.D., R.N., F.R.S.

DEPUTY INSPECTOR-GENERAL OF HOSPITALS AND FLEETS.
ASSISTANT PROFESSOR OF NAVAL HYGIENE, ARMY MEDICAL SCHOOL.

WITH TWENTY-FOUR LITHOGRAPHIC PLATES.

PHILADELPHIA
LINDSAY AND BLAKISTON
1875

LONDON:
SAVILL, EDWARDS AND CO., PRINTERS, CHANDOS STREET
COVENT GARDEN.

TO

SIR ALEXANDER ARMSTRONG, K.C.B., LL.D., F.R.S.

ETC. ETC. ETC.

DIRECTOR-GENERAL OF THE MEDICAL DEPARTMENT OF THE NAVY,

WHOSE NAME IS ESPECIALLY ASSOCIATED WITH THE CULTIVATION OF

THE SCIENCE OF HYGIENE IN HER MAJESTY'S NAVAL SERVICE,

This Work is Inscribed

WITH FEELINGS OF RESPECT AND ESTEEM,

BY HIS MOST OBEDIENT SERVANT,

THE AUTHOR.

PREFACE.

OFFICERS OF HEALTH, as well as Naval and Military Medical Officers, have often to determine the nature of the suspended matters in water used for drinking. In an Hygienic point of view, the import of these suspended matters must vary with their properties, whether mechanical, chemical, or vital.

Mineral particles may affect health, on account of their mechanical action, as for example, when mineral silt of clay, or fine sand causes diarrhœa. Dead animal and vegetable substances may have more important effects, as, when suspended fæcal matter produces irritation of the whole alimentary tract. On the other hand, living things, such as the ova of Entozoa, the nematoid worms, and small leeches may give rise at once to certain grave disorders, or Algæ may act on sulphates, and disengage sulphuretted hydrogen. There are, however, numerous living creatures, both animal and vegetable, found in drinking water, to which no special effect on health can be at present assigned; they may be important only as showing the presence of organic impurities, which serve as their pabulum, or as indicating putrefaction. Farther observation may, nevertheless, prove them to be of deeper sanitary significance, and even now, though there is no good evidence of their hurtful action, no one would hesitate to condemn a water containing Bacteria or fungi, or swarming with the lower forms of life. At any rate, whatever may be the conclusions hereafter arrived at, as to the sanitary import of the innumerable suspended matters, it cannot be doubted that Medical Officers of Health should be able to state what they are. This must be done chiefly by the microscope; but, as it is often difficult for those who are unacquainted with Natural History, even with a voluminous work of reference in their hands, to determine the nature of the various objects that may present themselves, the design of the following synopsis is to furnish a number of figures of those objects, with such a commentary as may enable them to be identified. No attempt has been made to link particular forms with special effects; it is doubtful indeed, if this be possible at present, beyond a limited extent, being rather a point for the enquiry of future times, which this little work can merely purport to aid.

The Tables and figures may also prove useful to young naturalists, who are beginning to investigate the world of waters, that wonderful world, in a single drop of which we may behold varieties of form, almost as numerous as those upon the surface of the great globe itself. Many books have been published with a similar object in view; but one more may find a place, to facilitate the study of a very interesting department of Natural Science.

In reference to the Plates, by way of apology, it may be mentioned that, with the view of lessening the expense of publication, the figures have been drawn with pen and ink, but, though they cannot pretend to the fineness and delicacy of steel engravings, some artistic effect has been preserved, and it is hoped that they will answer, equally well, the purpose for which they are intended.

To Professor Parkes, F.R.S., the thanks of the author are especially due, for his valuable advice and guidance, in rendering the treatment of the subject as practical as possible.

WESTON GROVE ROAD,
WOOLSTON, SOUTHAMPTON,
October 1*st*, 1875.

CONTENTS.

N.B.—*The numerals placed opposite the genera indicate the corresponding figures in the Plates.*

ERRATUM.

Page 42 *to end.*—(Plate XIV.) *should be* (Plate XV.), (Plate XV.) *should be* (Plate XVI.), and so on.

A GUIDE

TO THE

MICROSCOPICAL EXAMINATION

OF

DRINKING WATER.

INTRODUCTION.

Mode of Collecting Sediments and Placing them under the Microscope—Microscopical Powers—Immersion-lenses.

When water is very turbid, from an obviously impure source, it is easy enough to obtain a sufficient amount of sedimentary matter for microscopical examination, and a just estimate of the unfitness of such water for drinking purposes may be thus readily formed. But it more frequently happens that the deposit, even after long standing, is but slight, and when this is the case, we must have recourse to special means by which the whole, or a large amount of the matters in suspension may be concentrated, or collected together within a small compass. In the first place a tall glass vessel will be required, a litre or half-litre measure glass will answer very well, and when filled up to the mark, a circular disc of glass, resting upon a horizontal loop at the end of a long wire, should be let down to the bottom of the vessel, and the whole arrangement, lightly covered, set aside for twenty-four or forty-eight hours, as the case may be.

At the end of the specified time, the water may be siphoned off with a piece of india-rubber tubing, only leaving a thin stratum over the glass disc. This should now be carefully raised, and laid upon a piece of blotting-paper, so as to dry its under surface, when it may be at once transferred to the microscope, with a large piece of thin covering glass so placed upon it as to exclude all air-bubbles.

Another good plan is to siphon off the water until only a sufficient quantity remains to permit the sediment to be shaken up with it, and poured into a tall conical glass, from which, after standing again for a short time, portions may be taken up by means of a pipette, and placed on a slide for examination. A thin glass cover is always required, not only to equalize the refraction, but to confine the fluid and prevent evaporation, by which an obstructive dew would naturally form upon the object glass. In the manner just mentioned the specimen of well-water sediment represented in Plate XXIII. was prepared. The gelatinous matter, developed by the bacteria-like cells at the lower part of the figure, (*a*) was only loosely adherent at the close of the first day; but, subsequently to this, or during the next forty-eight hours, it formed a delicate but perfect incrustation at the bottom of vessel. Many of the little bodies, detached from the gelatinous frond, were seen in active motion in the immediate vicinity. More definite fronds, with still more minute bacteriform bodies growing upon a decomposing spray of pond weed, are shown in Plate VII. as seen with a sixteenth of an inch immersion-lens. The first of these forms, at least, would seem to exhibit an alliance with the *Palmellaceæ*, while others, which are very readily confounded with them, show a marked affinity with the *Oscillatorians* (see further remarks on this subject under the head of *Bacteriaceæ*).

It will be apparent, from the foregoing observations, that the sediments of comparatively clear water require the very highest microscopical powers for their investigation, and the employment of immersion-lenses if available. Filamentous algæ, even narrower than true bacteria, may be thus frequently brought into view, as well as the delicate flagellæ or locomotive organs of monads, whose bodies alone would be scarcely visible with lower powers. It is also important to mention that, by these means, even in the absence of ordinary amœbæ, particles of protoplasm of bacterium size, exhibiting amœbiform movements, are often discernible. Lastly, very finely-divided mineral matter in suspension, giving rise to milkiness or haze, can only be studied with immersion-lenses, though certain cases may occur in which no objective cause of these conditions can be detected microscopically.

Mineral matters of various hues in the soil, through or over which water percolates or flows, are the more usual causes of discoloration and turbidity. Peroxide of iron, in particular, may be mentioned as the source of the brown cloudy appearance of water from the blue clay, as also, frequently, of the brown colour of pools in bog-lands, though this is more likely to arise from organic matter in a state of decay. In the coarser sediment, under such circumstances, the microscopic forms of animal and vegetable life are likely to be abundant (viz., *Rotifera* and *Infusoria*, *Oscillatoriaceæ* and *Desmidiaceæ*). In ferruginous bog-water also the twin-spiral filaments of *Didymohelix*, invested with a yellowish-brown gelatinous matter, and *Leptothrix ochrea*, a rather ill-defined mycelioid structure, may add to the general effect. By reflected light, moreover, the fine amber tint of the *Diatomaceæ*, floating or resting, is quite brown. Some of the heterogeneous materials usually occurring in bog-water are represented in Plate XXIV.

SECTION I.

Mineral Matters. (Plate I.)

Mineral matters in suspension in water often give a turbidity of a colour and character indicative of their nature. When the particles are large, they will descend more rapidly; but when very subtle or minutely divided, the suspension being more complete, a longer time will be required for their subsidence. Looking down through a considerable depth of the water, with the glass vessel containing it resting on a white ground. will afford some preparatory information, when compared with a corresponding stratum of distilled water in a second vessel observed in a similar way. Haziness, or peculiarity of colour, may be thus detected, which would be quite inappreciable in a thin layer. With a long glass tube a stratum of two or more feet might be obtained, and the method is also valuable in observing the effect of reagents or tests in water. In the light of preparatory information, it may be stated, moreover, that sandy particles and clay in suspension give a yellowish-white turbidity; and on boiling the water, as Professor Parkes observes, "sand, chalk, and heavy particles of the kind will be deposited," and if it be a chalk water the calcium carbonate will carry down suspended sewage or vegetable matter, effecting a change of colour. Under such circumstances the sense of smell may afford confirmatory evidence.

Silicious particles, as of flint or sand, are usually angular; and though often much rounded by rolling and attrition, vitreous fracture will be observable in many of them, as shown in Plate XXIII. It should be mentioned here, that a little source of fallacy may be occasioned by the very frequent detachment of minute scales or chips from the margins of the covering glass, or the extremity of the pipette, when not properly ground, or even from the glass stoppers of bottles in which specimens are kept. On carefully inspecting the more minute particles of silicious matter, which are so easily diffused and suspended in water, their thin or scale-like character will be apparent. Particles of chalk, clay, and marl, on the other hand, are usually more rounded, but the former will be at once recognised by their solubility in acids. The crystalline forms of numerous substances are frequently visible in the smallest molecules. Indeed, the study of the inorganic matters in the sediments of fresh water, is a branch of Microscopical Mineralogy which is of growing interest and importance to the water analyst. It would of course include goniometry and spectrum analysis, and will no doubt receive the attention it deserves in time to come.

SECTION II.

Dead, or Decaying Organic Matter.

Any of the forms described in the succeeding Section, as living plants and animals, may be found in the sediment of drinking water, either whole or fragmentary, in a dead and more or less decayed state. Their recognition will, in many cases, be difficult in consequence of the accumulation of débris of different kinds about them, as well as their own altered condition. But, when the more unyielding structures remain intact, a little practice, with the help of figures, will enable the observer to determine them with sufficient accuracy for all practical purposes.

A. *Dead Vegetable Matter.* (Plates II. & III.)

When the higher plants die down, those of a more humble kind seem to flourish with greater vigour, so that however shapeless the decaying masses may be, minute *Oscillatorians*, *Bacteria*, and their allies will usually be found in their vicinity. The breaking down of vegetable cells is of course attended with the discharge of the contained cell-sap, endochrome, &c.; and these will soon assume an amorphous, or irregular granular appearance, in which the original green colour is here and there very evident. Its further change, however, is usually into an olivaceous or yellowish-brown tint. In some instances the albuminous inner coat of the vegetable cells, known as the primordial utricle, is seen much contracted within the cellulose coat, passing into an indigo purple tint in a more advanced stage of decay. With a little care, the collapsed and crumpled cell walls may be recognised casually. But, very characteristic of decaying vegetable matter, if it appertain to vascular plants, is the occurrence of spiral vessels, or even the spiral fibres drawn out of the cells; annular ducts, dotted and pitted tissue, and hairs, which, from their comparative indestructibility, are sometimes very beautifully dissected out, as it were, by maceration. These at once afford a clue to the nature of the amorphous matter in connexion with which they are found.

The little scales of "bog moss" (*Sphagnum*), with their porous or fenestrated cells, the discs and roots of duckweed (*Lemna*), and sprays of "pond weed" (*Potamogeton*), and the "stoneworts" (*Chara* and *Nitella*), may also be met with, more or less altered in colour, or otherwise.

Amongst the vegetable products (Plate III.) not properly belonging to the fresh water, but indicating contamination from house refuse, may be mentioned the fibres of fabrics, such as linen (1), the hemp of twine (2), cotton (3), and the discoidal tissue of ordinary deal or pine (4), a structure, it may be remarked, which is characteristic of the *Coniferæ* as a whole.

B. *Dead Animal Matter.* (Plates IV. & V.)

Decaying animal, as well as vegetable matter, may consist of materials proper to the fresh water or foreign to it. To the first class belong, in particular, the dead bodies of the *Entomostraca* (water fleas, &c.), and the numerous forms of segmented or *Annulose* animals, including the water-bears and mites, the larvæ of aquatic insects, and the *Annelida*. Indeed the latter are often only to be recognised by their indestructible setæ and ventral hooks, which may ultimately become quite isolated in the field. Animal products, not proper to the fresh water, may embrace the bodies or exuviæ of terrestrial insects, houseflies and others, often overgrown with *Achlya*, a parasitic siphonaceous plant (Plate XII. 5), and matters such as are represented in Plate V., to which the following references will apply:—

1. Fibres of silk. 2. Wool. 3. Human hair. 4. Hair of rabbit—*a*, the shaft, and *b*, the extremity. 5. Epithelium from the mouth. 6. Ditto from the cutaneous surface. 7. Striped muscular fibre. 8. A feather. 9. Portions of ditto, more highly magnified. 10. Scales of Lepidoptera.

The scales of moths and butterflies are usually flat, with fine longitudinal fluting and a serrated extremity. Hairs properly so called have commonly a soft central axis of cells, often absorbed so as to form a medullary cavity. Wool, on the other hand, is much smaller and more compact in the centre, while the superficial imbrication of the component cells is more distinctly marked. Human epithelial scales are broad and flat, with an oval highly refringent nucleus and minute scattered points in the surrounding space. They resist maceration for a considerable time, and thus frequently percolate with other impurities from latrines into neighbouring wells.

It may not be out of place here to call the attention of the observer to the possible presence of the eggs of Entozoa in the water under examination. All spherical and ovoid bodies with albuminous-looking and segmented contents should be looked upon with suspicion, until their real nature has been determined; accurate measurements of them should be taken, and drawings if possible. (See Plate XXII., Figs. 10—14.)

SECTION III.

Living Forms.

The simplest grades of plants and animals or the Protophyta and Protozoa possess so many characters in common, that it is by no means easy to determine the true nature or position of numerous minute organisms, which constantly present themselves in the field of the microscope. The most reliable means of distinguishing them is founded upon physiological grounds, and more especially their mode of nutrition. For it is quite admitted that no structural particulars can be named, in the abstract, as characterizing the one more than the other. Of course, where the life history of any form has been satisfactorily traced out the determination must be certain; as for example, when a Zoospore, furnished with motile organs or flagella, is found not only to originate in a bonâ-fide plant, but ultimately to grow into one itself. Of such organisms, unquestionably, Dujardin's *Flagellata,* or first Order of Infusoria (B, I. 2) (*a*) mainly, if not altogether, consists. Others of a similar description usually associated in groups in a gelatinous frond, occur in the *Volvocaceæ* (A, II. 7). To the casual observer, the equivocal movements executed by the forms of doubtful position are more striking than their intimate structure, while the other parts of their history are quite out of the question. Indeed, in many cases, a claim to belong to the animal kingdom has been raised alone upon the exhibition of animal-like movements. The liability to error is therefore all on one side, and as far as we know not a single genuine protozoon has ever been classed by the botanist in his domain, while our greatest difficulty at the present time is to eliminate the protophyta from the realm of zoology.

It will be scarcely doubted that the numerous species of *Difflugia, Arcella,* and *Euglypha* are veritable animals; but what are we to say of the equally numerous *Amœbæ,* now that we are acquainted with the truly vegetable *Amœboids* of *Volvox,* and of the roots of mosses, through the researches of Dr. Hicks, F.R.S. The pliant *Vibrio* and the rigid *Diatom* exhibit the phenomenon of spontaneous movement, connected no doubt with the play of the same, or similar nutritive processes, developing dialytic currents, which are on this account quite invisible, while they operate as a moving cause on moveable bodies. In this way the rigid diatom moves without change of form, and shall we say by the same law the extensile plasma of the passive amœboid is drawn out into pseudopodia, with the semblance of active, and even of voluntary motion?

The following kinds of movement may be noticed and compared in the two kingdoms:—

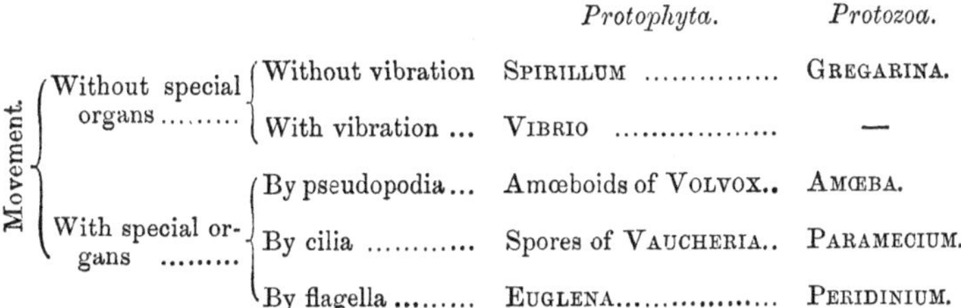

			Protophyta.	*Protozoa.*
Movement.	Without special organs	Without vibration	SPIRILLUM	GREGARINA.
		With vibration ...	VIBRIO	—
	With special organs	By pseudopodia...	Amœboids of VOLVOX..	AMŒBA.
		By cilia	Spores of VAUCHERIA..	PARAMECIUM.
		By flagella	EUGLENA..................	PERIDINIUM.

Above the lowest grades of plants and animals, or such as are notified in the preceding table, no difficulty can arise in assigning to every form its true position.

A. *Living Plants.*

COMPRISING THE MORE USUAL AQUATIC ALGÆ OCCURRING IN THE EXAMINATION OF DRINKING WATER.

Though our knowledge of the fresh-water Algæ has become greatly extended of late years, we are still only in possession of fragmentary particulars in relation to many of the more humble forms; and until the whole life-history of each has been satisfactorily traced out, it would be quite impossible to group them so as to be altogether free from objection. The classification here adopted cannot, therefore, purport to be perfect, but it is hoped that it may serve as a guide to the leading characters of the vegetable products usually presented to the observer in the microscopical examination of drinking water.

Systematic Arrangement.

The numerous types of fresh-water Algæ would appear to admit of natural distribution into three groups or sections, distinguished as follows:—

Group I. Plants which, although for the most part exhibiting spontaneous motion in themselves, have yet no special provision for movement in their reproductive elements.

Families included in this group, viz.:—

1. *Bacteriaceæ.* (*Bacteria,* of Cohn.)
2. *Oscillatoriaceæ.*
3. *Nostochaceæ.*
4. *Palmellaceæ.*
5. *Desmidiaceæ.*
6. *Diatomaceæ.*

Group II. Plants in which motile (*i.e.,* ciliated or flagellate) cells play the most conspicuous part, either separately, simply aggregated, or organically united in a definite manner in a gelatinous frond.

This group includes a single family, viz. :—

7. *Volvocaceæ.*

Group III. Plants in which all movement is confined to the reproductive elements, comprising the remaining families, viz. :—

8. *Pediastraceæ.*
9. *Ulvaceæ.*
10. *Apiocystaceæ.*
11. *Siphonaceæ.*
12. *Zygnemaceæ.*
13. *Confervaceæ.*
14. *Œdogoniaceæ.*
15. *Chætophoraceæ.*
16. *Batrachospermaceæ.*
17. *Characeæ.*

Definition of the foregoing Families, and of the more important Genera appertaining to them.

GROUP I.

Family I.—*Bacteriaceæ.* (Plates VI. & VII.)

Under the head of *Bacteria,* Cohn has included all the very minute spherical, elongated, rod-like, straight, and spiral filamentous plants endowed with more or less active spontaneous motion; and now found to be associated with putrefaction and other conditions of hygienic importance.

The annexed table is in accordance with Dr. Cohn's own classification, which he admits must be only provisional, until something more definite is known of the nature and affinities of these interesting organisms. Though the species are not separately described, it was considered advisable to retain them in the table to facilitate further reference, should it be found necessary.

Bacteria (Cohn).

			GENUS AND SPECIES.
A. SPHÆROBACTERIA.			*Micrococcus.*
(Minute jostling spherules.)	Zymogenous. (Ferment producing.)	(*c*)	crepusculum (Ehr.)
			candidus (Cohn)
	,,	(*d*)	ureæ (Cohn) The ferment of ammoniacal putrescence.
	Chromogenous. (Colour producing.)	(*a*)	prodigiosus (Ehr.) The blood stain in bread.
	,,		luteus (Schrœter)
	,,		aurantiacus (Sch.)
	,,		chlorinus (Sch.)
	,,		cyanus (Sch.)
	,,		violaceus (Sch.)
	Pathogenous. (Disease producing.)	(*b*)	vaccinæ (Cohn)
			diphthericus (Dartel)
	,,		septicus (Klebs)
	,,		bombycis (Béchamp)
B. MICROBACTERIA.			*Bacterium.*
(Minute and short rods.)		(*f*, *g*, *h*, *k*)	termo (Ehr.) Producing putrefactive fermentation.
		(*i*, *l*)	lineola (Ehr.) In brooks, &c.
	Chromogenous.		xanthium (Sch.)
	,,		syncyanum (Sch.)
	,,		æruginosum (Sch.)
C. DESMOBACTERIA			*Bacillus.*
(Straight, flexible or rapidly undulating filaments.)		(*n*)	subtilis (Ehr.) Producing Butyric fermentation.
		(*m*)	ulna (Kohn) Similar to the former.
			anthracis (Cohn) In the blood, in malignant pustule.
D. SPIROBACTERIA.			*Vibrio.*
(Spiral filaments, rigid, or flexible.)		(*o*)	rugula (Ehr.)
		(*p*)	serpens (Ehr.)
			Spirillum.
		(*q*)	tenue (Ehr.)
		(*r*)	undula (Cohn)
		(*s*)	votutans (Ehr.)
			Spirochæta.
		(*t*)	plicatilis (Ehr.)

While there is little doubt of the intimate relationship existing between the larger forms of the preceding table and the Oscillatorians, *Bacterium termo* and its immediate allies are involved in much obscurity as to their real nature and botanical affinities, seeing that their supposed position in the animal kingdom is now no longer tenable. The slightly dumb-bell shape of the true putrefactive *Bacterium* manifests a very significant correspondence with the form represented in Plate XXIII., developed in the sediment of well-water, and with many others such as that shown in Plate VII., occurring amongst decomposing Algæ.

All analogy would go to indicate that the Zoogl[illegible]ea form of *Bacterium termo* may be regarded as the primary or normal state of this organism, the surrounding gelatinous matter being simply the representative of that which forms the indefinite frond of *Microhaloa* or *Palmella* for example.

Further, when the matrix breaks down, and the separate little *Bacteria* detach themselves from it, they often commence those active movements which are in some intimate way connected with their nutrition. Even many *Diatomaceæ* which are normally fixed to, or included in a gelatinous frond are motionless until they have become free from it, when the movements they exhibit are known to bear a certain relation to the shape of the frustule, being rectilinear when the latter is narrow, but more irregular when it is of a different form. The subsequent history of *Bacteria* has been variously represented by authors, but our space will not admit of further enlargement upon this subject.

The carbon of the higher aquatic plants is derived from the carbonic acid present in the water, or liberated by the decomposition of carbonates, while that of the molecular and more minute filamentous Algæ (*Micrococcus, Bacterium*, &c.) is usually derived from the vegetable acids that may be in combination with a base, as for example, the T̄ of Tartrate of Ammonia.

Dr. Cohn's researches go to show that, not only will *Bacteria* flourish in solutions of the salt just mentioned, or in the absence of organic matter, but that even in this case the genuine putrefactive odour is evolved. This important fact would therefore link the presence of *Bacteria* with putrefaction as a process quite distinct from simple decay, with which fungous-life is more particularly associated.

Family II.—*Oscillatoriaceæ*. (Plate VIII.)

These very simple plants consist of tubular filaments, with or without a gelatinous investment, and having faint or rich bluish-green or purple coloured contents, or endochrome, in which, as the filaments elongate, a transverse segmentation takes place, giving rise to the deceptive appearance of cells in single series. The filaments may be quite free, or disposed in bundles or strata. In the free state, their peculiar animal-like movements render them objects of interest to the microscopist. Branching, in the true sense of the term, is quite foreign to these plants, which multiply by transverse fission; but of their sexual reproduction nothing is precisely known. Excluding the *Bacteria* of Cohn, they are divided into several sub-families, which are easily distinguished in the following manner:—

				Sub-families.		*Genera.*
Frond or filament.	Cylindrical.	Exhibiting more or less active movements		*Oscillatorieæ*...	(1.)	Oscillatoria.
		Motionless	Like Oscillatoria, but in tufts or strata.	*Lyngbyeæ*......	(3.)	Lyngbya.
			Having a proper gelatinous investment	*Scytonemeæ* ...	(4.)	Scytonema.
				....	(2.)	Microcoleus.
	Tapering, with a large basal cell......			*Rivularieæ* ...	(5.)	Rivularia.

It is highly probable that many of the supposed members of *Oscillatoriaceæ* are truly referable to the succeeding family (*Nostochaceæ*).

On carefully inspecting a fair specimen of water rich in Algæ of different kinds, it will usually be easy to trace examples of *Oscillatorieæ* ranging from the proportions of ordinary *Confervæ* to the diameter of *Bacterium termo*. The same phenomena of endochrome-cleavage and spontaneous movement will be seen to occur in all cases, in a more or less marked degree; and indeed any differences distinguishable in the smaller, as compared with the larger forms, can only be said to be of a relative kind, and apparently in no way contraindicate a prevailing unity of type. Frequently also the smallest moving points or molecules observable in the field, instead of being referable to the genus *Micrococcus*, are but segments of the more minute filamentary species or varieties, as the case may be; for, even if their cylindrical form is not demonstrable to the eye, their peculiar refractive properties will enable us to link them with the less equivocal fragments always to be found in the same vicinity.

In the punctiform, fragmentary, or filamentous plants of smaller size than the admitted Oscillatorians, it is impossible to distinguish a primordial utricle and a cellulose coat, and of course also difficult to determine the precise nature of the segmentation. In the *Oscillatorieæ*, however, the endochrome suffers cleavage, while the primordial utricle and the cellulose tube take no part in the process, being only capable of simple growth and extension. In the *Confervaceæ* and other filamentous Algæ, on the contrary, both the primordial utricle and the endochrome are engaged in the segmentation of the filament, within the cellulose coat, to which, nevertheless, the transverse septa and a new internal layer are added.

In a very interesting paper published in the *Quarterly Journal of Microscopical Science,* vol. i. 1861, Dr. B. Hicks, F.R.S., has touched a most important subject in what he has termed the Diamorphosis of *Lyngbya muralis.* This plant, though confounded by some with the genuine *Confervæ,* is now generally admitted to be an ally of the Oscillatorians, and as such at least one of its modes of reproduction, or transitional phases, presents a suggestive bearing upon all the members of this family, and thereby, it may be fairly presumed, upon *Bacteriaceæ* in general. We thus perceive how slender are the grounds upon which we can assume almost any palmellaceous plant to be a distinct entity, and in this remark may be included some forms reputedly belonging to the *Ulvaceæ.* Let us suppose for a moment that the minute spirilla and even *Bacterium termo* itself are in the category of the filamentous algæ, then how small must be their reproductive gonidia!

From actual observation of the spirillum common in bilge-water, I can safely say that the moving particles in which it originates, however small they may have been in the first instance, are not only very minute, but quite shapeless. If these reproductive particles are visibly so small in relation to the diameter of a normal filament of *Lyngbya muralis,* how minute must they be in the case of *Bacterium termo!* They might readily escape the keenest scrutiny of the advocates of equivocal generation.

FAMILY III.—*Nostochaceæ.* (PLATE VIII.)

Plants consisting of microscopic moniliform filaments of cells in series, usually coiled, curved, or entangled in a gelatinous matrix constituting the frond, which may be round or foliaceous, linear, or formless. They are found on damp ground, or in water, floating on the top, or at the bottom, attached to stones in rivulets and streams, or in brackish ditches.

The characters of the frond sufficiently distinguish the three more important genera, thus—

Frond	*Expanded* ...	Globular or irregular; filaments numerous .	(1.)	NOSTOC.
	Elongated ...	Curved, linear, or spiral; filaments single .		MONORMIA.
	Formless ...	Often a floating film; filaments numerous .	(2.)	TRICHORMUS.

Besides simple multiplication by fission (which is sometimes longitudinal as well as transverse), the *Nostochaceæ* afford indications of the existence of a true reproductive process, in the presence of certain vesicular cells (supposed to be spermatic?) amongst the ordinary ones; which latter are, moreover, here and there further developed into sporangial cells, producing true spores from which new filaments arise. This process appears to have been distinctly observed by Thuret in *Nostoc verrucosum*.

The three remaining families of this section are, strictly speaking, composed of unicellular plants—*i.e.*, consisting essentially of a single cell, which may be solitary or associated with others, in no very definite order, or as a brittle filament; cells multiplying by fission and reproducing by conjugation.

FAMILY IV.—*Palmellaceæ*. (PLATE IX.)

Green cells (often red), spherical or ovate, in a more or less consistent or definitely formed gelatinous material, constituting a frond, so called; the cells multiplying by simple fission, without gemmation. Of the numerous genera referred to this family, the following may be given as good examples—

Frond	Indefinite or formless......	Mucoid, *floating*, with minute cells	(1.) MICROHALOA.
		Slimy, *encrusting*, with large globular cells	(2.) PALMELLA.
	More consistent and definite in form.	Globular, including numerous distinct cells	(3.) COCCOCHLORIS.
		Bandlike, simple, or branched, with cells in twos or fours in single series	(4.) HORMOSPORA.

Though the precise limits of the *Palmellaceæ* are yet but imperfectly defined, these plants are of considerable interest to the water analyst, they so frequently find their way into cisterns and reservoirs, and thus make their appearance in the deposits of drinking water. Several genera which would appear to be more correctly referrible to the *Volvocaceæ*, are usually confounded with them; and the accumulation of synonyms has only added to the confusion.

To illustrate multiplication by fission in the *Palmellaceæ*, we shall instance the genus *Coccochloris*, which will enable us to see what little more is required to meet the conditions observable in the *Desmidiaceæ* and *Diatomaceæ* respectively.

In *Coccochloris* (3 *a* and *b*) binary subdivision, with the successive formation of a cellulose and hyaline investment, seems to go on practically without limit, a fresh impetus to the process being given by the conjugation and blending of two endochromes (*c*), in which repeated fission goes forward as before. This is, in effect, also what takes place in the *Desmidiaceæ* and *Diatomaceæ*, and the observation is so far correct, even though *Coccochloris* and its allies should be, as some suppose, but the gonidia of Lichens in a certain phase of development.

FAMILY V.—*Desmidiaceæ*. (PLATE IX.)

These are unicellular plants, usually of an exceedingly rich green colour, nearly exclusively confined to fresh water, occurring singly, or remaining in contact after binary subdivision, so as to form more or less brittle threads of cells in linear series. A sutural line running round the cell-wall transversely, marks it off into two symmetrical halves, and cleavage takes place at this line, preparatory to the gemmation of two new half frustules from the old ones thus separated. The forms of these cells are very beautiful and varied, and chiefly characterize the genera, which admit of the following arrangement:—

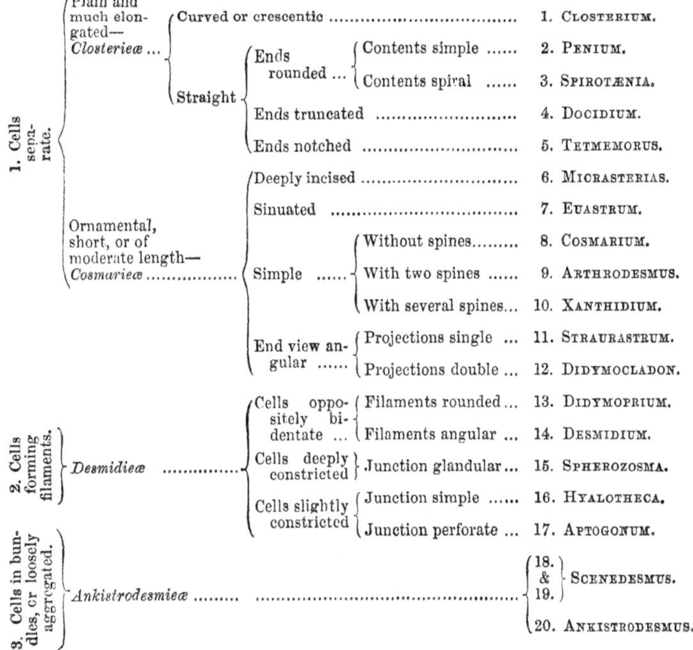

1. Cells separate.	Plain and much elongated—*Closterieæ* ...	Curved or crescentic ...			1. Closterium.
		Straight	Ends rounded ...	Contents simple	2. Penium.
				Contents spiral	3. Spirotænia.
			Ends truncated ...		4. Docidium.
			Ends notched ...		5. Tetmemorus.
	Ornamental, short, or of moderate length—*Cosmarieæ* ...	Deeply incised ...			6. Micrasterias.
		Sinuated ...			7. Euastrum.
		Simple		Without spines.........	8. Cosmarium.
				With two spines	9. Arthrodesmus.
				With several spines...	10. Xanthidium.
		End view angular		Projections single ...	11. Straurastrum.
				Projections double ...	12. Didymocladon.
2. Cells forming filaments.	*Desmidieæ* ...	Cells oppositely bidentate ...		Filaments rounded ...	13. Didymoprium.
				Filaments angular ...	14. Desmidium.
		Cells deeply constricted		Junction glandular...	15. Spherozosma.
		Cells slightly constricted		Junction simple	16. Hyalotheca.
				Junction perforate ...	17. Aptogonum.
3. Cells in bundles, or loosely aggregated.	*Ankistrodesmieæ* ...				18. & 19. Scenedesmus.
					20. Ankistrodesmus.

Family VI.—*Diatomaceæ*. (Plate X.)

Like the former family, the *Diatomaceæ* are unicellular plants, in some instances isolated, in others cohering in chains or filaments, or in some definite way. The cell wall, however, is composed of a glassy or silicious material, instead of cellulose, which is found in all other vegetable cells; and the endochrome is usually of a rich amber tint instead of green. They exhibit also much symmetry and beauty in the forms of the frustules, which are often so exquisitely sculptured as to afford excellent test objects for the microscope.

Each frustule consists of a new and an old half or valve, as noticed in the *Desmidiaceæ*, but the margins of the old valve overlap those of the new one, and thus results the so-called cingulum or "middle piece," which is not only capable of elongation by growth, but also by one portion sliding upon the other, telescope fashion, so as to make provision for the endogenous development of two new half frustules by fission and gemmation combined. From this arrangement it follows that the cells of each successive generation must be narrower than those within which they arise, by at least the whole thickness of the cell wall. Here then is the explanation of the great disparity of size so frequently observed in members of the same species. Moreover, we thus also see why it is that after the conjugation of two frustules, the resulting sporangial cell, in which the process just described commences, should be so much larger than the parent cells.

The genera of *Diatomaceæ* are too numerous to be separately defined in this treatise; but the annexed table, with the figures arranged in the same order, will assist in the recognition of the more usual fresh water forms :—

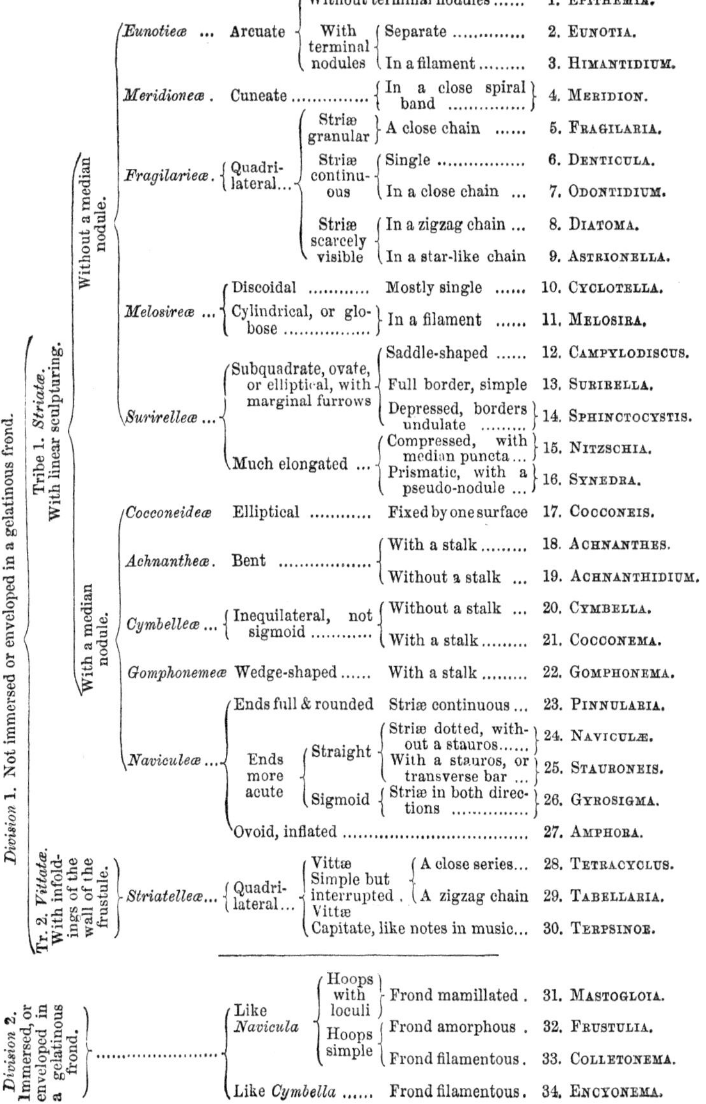

Division 1. Not immersed or enveloped in a gelatinous frond.	Tribe 1. *Striatæ.* With linear sculpturing.	Without a median nodule.	*Eunotieæ* ...	Arcuate	Without terminal nodules		1. EPITHEMIA.
					With terminal nodules	Separate	2. EUNOTIA.
						In a filament	3. HIMANTIDIUM.
			Meridioneæ .	Cuneate		In a close spiral band	4. MERIDION.
			Fragilarieæ .	Quadri-lateral...	Striæ granular	A close chain	5. FRAGILARIA.
					Striæ continuous	Single	6. DENTICULA.
						In a close chain ...	7. ODONTIDIUM.
					Striæ scarcely visible	In a zigzag chain ...	8. DIATOMA.
						In a star-like chain	9. ASTRIONELLA.
			Melosireæ ...	Discoidal		Mostly single	10. CYCLOTELLA.
				Cylindrical, or globose		In a filament	11. MELOSIRA.
			Surirelleæ ...	Subquadrate, ovate, or elliptical, with marginal furrows		Saddle-shaped	12. CAMPYLODISCUS.
						Full border, simple	13. SURIRELLA.
						Depressed, borders undulate	14. SPHINCTOCYSTIS.
				Much elongated ...		Compressed, with median puncta ...	15. NITZSCHIA.
						Prismatic, with a pseudo-nodule ...	16. SYNEDRA.
		With a median nodule.	*Cocconeideæ*	Elliptical		Fixed by one surface	17. COCCONEIS.
			Achnantheæ.	Bent		With a stalk	18. ACHNANTHES.
						Without a stalk ...	19. ACHNANTHIDIUM.
			Cymbelleæ ...	Inequilateral, not sigmoid		Without a stalk ...	20. CYMBELLA.
						With a stalk	21. COCCONEMA.
			Gomphonemeæ	Wedge-shaped		With a stalk	22. GOMPHONEMA.
			Naviculeæ ...	Ends full & rounded		Striæ continuous ...	23. PINNULARIA.
				Ends more acute	Straight	Striæ dotted, without a stauros......	24. NAVICULÆ.
						With a stauros, or transverse bar ...	25. STAURONEIS.
					Sigmoid	Striæ in both directions	26. GYROSIGMA.
				Ovoid, inflated			27. AMPHORA.
	Tr. 2. *Vittatæ.* With infoldings of the wall of the frustule.		*Striatelleæ*...	Quadri-lateral...	Vittæ Simple but interrupted . Vittæ	A close series...	28. TETRACYCLUS.
						A zigzag chain	29. TABELLARIA.
					Capitate, like notes in music...		30. TERPSINOE.
Division 2. Immersed, or enveloped in a gelatinous frond.			....	Like *Navicula*	Hoops with loculi	Frond mamillated .	31. MASTOGLOIA.
					Hoops simple	Frond amorphous .	32. FRUSTULIA.
						Frond filamentous.	33. COLLETONEMA.
				Like *Cymbella*		Frond filamentous.	34. ENCYONEMA.

GROUP II.

FAMILY VII.—*Volvocaceæ.* (PLATE XI.)

THIS family is sufficiently defined in the terms of Section II., and it will only be necessary to characterize the leading genera.

Motile or flagellate cells.	Single, two, or four in number, remaining united by incomplete cleavage			(1.) PROTOCOCCUS.
	Numerous.	Grouped by fours in larger cells on a gelatinous frond		(2.) TETRASPORA.
		Free, evenly distributed in a round or oval frond		(4.) PANDORINA.
		Mutually united by stolons, but originally distinct	In square tablets ...	(3.) GONIUM.
			In spherical extensions	(5.) VOLVOX.

The life history of *Protococcus,* so far as it has been traced out by Cohn and others, presents such a variety of conditions and stages that it is difficult to retain them in the memory. It presents, however, so close a relationship to *Volvox,* that it would be well to compare the two forms carefully and contrast both with *Pediastrum* and *Hydrodictyon.*

In one developmental stage of *Protococcus,* a motile cell encysted after a fashion, breaks up into four by cleavage, but frequently these remain united by their beaked extremity, when the cleavage has not been quite completed, thus producing a compound form strikingly suggestive of *Volvox,* which is simply a wider extension of a similar condition.

The Volvox sphere results from the segmentation of a single mass of endochrome, the ultimate subdivisions of which assume the flagellate motile character, and become organically united by the mutual blending of little stolon-like extensions, piercing the hyaline investments, which become hexagonal by mutual contact and compression. This union also takes place in the cells of *Gonium*, while in numerous other *Volvocaceæ*, as in *Pandorina*, it never happens, but the cells simply lie in juxtaposition. As before intimated, the connexion of the four motile cells of *Protococcus* arises from the incomplete cleavage of the original cell, while the communication subsisting between the cells of *Volvox* and *Gonium* is sequential to complete cleavage.

The union of primary distinct elements to constitute what we must regard as the perfect organism is further seen in the *Pediastreæ*, which are at present, obviously incorrectly, associated with the *Desmidiaceæ*, and in the remarkable genus *Hydrodictyon*, supposed to be siphonaceous. These are provisionally arranged by themselves in the next Section.

GROUP III.

FAMILY VIII.—*Pediastraceæ*. (PLATE XI.)

As defined in the preceding paragraph, including at least three genera.

1. Cells disposed in radiate discoidal fronds always minute .... (1.) *Pediastrum.*
2. Cells like those of a Pediastrum, but in a spherical frond .... *Cœlastrum.*
3. Cells disposed in a reticulate sacculus, often attaining a considerable size... (2.) *Hydrodictyon.*

In *Pediastrum* the form originates in the cleavage of an endochrome into two, then four, and finally some multiple of this, when a radial frondose expansion is formed by the juxtaposition, and union of the cells in some definite manner.

In *Hydrodictyon,* on the other hand, a motile cell breaks up into numerous distinct endochromes, which acquire a cellulose coat, and so arrange themselves as to form a reticulatiou of minute cylindrical cells, which gradually increase in size, and finally attain the character and dimensions they exhibit in the perfect plant. *Hydrodictyon* would therefore appear to hold a relationship to *Pediastrum,* similar to that which *Volvox* bears to *Gonium* or *Protococcus.* The latter organisms being made up of motile, and the former of ordinary vegetable cells.

Family IX.—*Ulvaceæ.*

Plants composed of a single or double layer of green polyhedral cells, multiplying by fission, disposed in tabular or tubular frondose extensions, chiefly marine, but in some few instances occurring in brackish or fresh water.

The long tubular fronds of *Enteromorpha intestinalis* are sometimes found in fresh-water ditches, but perhaps more usually in brackish or salt water.

Family X.—*Apiocystaceæ.* *Siphonoid* (unicellular) *Algæ.*

(Plate XI.)

The members of this family seem to be grouped with the *Palmellaceæ* as a matter of convenience. They are, however, quite distinct in their habits and relations. The fronds are composed of single cells, usually fixed at one end, and the reproductive elements are developed in the same cells, apparently engaging their whole contents. The following genera will serve for illustration :—

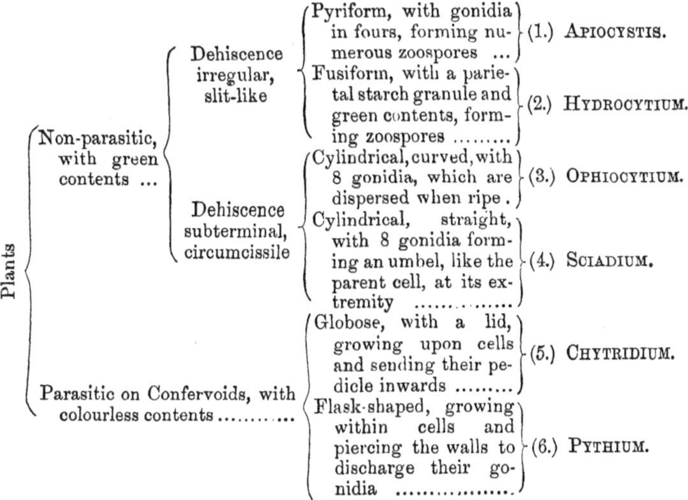

Plants	Non-parasitic, with green contents ...	Dehiscence irregular, slit-like	Pyriform, with gonidia in fours, forming numerous zoospores ...	(1.) APIOCYSTIS.
			Fusiform, with a parietal starch granule and green contents, forming zoospores	(2.) HYDROCYTIUM.
		Dehiscence subterminal, circumcissile	Cylindrical, curved, with 8 gonidia, which are dispersed when ripe .	(3.) OPHIOCYTIUM.
			Cylindrical, straight, with 8 gonidia forming an umbel, like the parent cell, at its extremity	(4.) SCIADIUM.
	Parasitic on Confervoids, with colourless contents		Globose, with a lid, growing upon cells and sending their pedicle inwards	(5.) CHYTRIDIUM.
			Flask-shaped, growing within cells and piercing the walls to discharge their gonidia	(6.) PYTHIUM.

FAMILY XI.—*Siphonaceæ*. (PLATE XII.)

Fronds unicellular, or composed of a continuous extension of simple membrane, with the reproductive elements developed in special organs or cells.

Excluding such members of this family as are purely marine, only two fresh-water genera are worthy of special notice here—viz., *Vaucheria* and *Achlya.*

(1 to 4.) *Vaucheria.* Most of the species of this genus are inhabitants of fresh water; but some are marine. They consist of branched tubular filaments, frequently almost felted together in fine silky green tufts. The little granules of chlorophyll in the interior of the filaments are for the most part applied to the walls, embedded in a colourless protoplasm. *Zoospores* are formed in the club-shaped ends of the filaments. Unger observed that these bodies usually made their escape about eight o'clock A.M., at which time the process may be

observed in healthy plants cultivated in fresh water. A true sexual mode of reproduction also exists in *Vaucheria*. Of the numerous species of this genus that have been described it would appear that only two or three are reliable. 2, portion more highly magnified; 3, reproductive organs; 4, *a*, *b*, and *c*, stages of development of the ciliated spore.

(5.) *Achlya prolifera* is a small colourless plant, consisting of clavate erect tubular filaments springing from a mycelium-like minutely ramified base, closely applied to the bodies of dead flies in water, fish and frogs, upon which they grow parasitically. It was originally mistaken for the common fly fungus, or an aquatic form of *Botrytis Bassiana*, but more recent researches, rewarded by the discovery of ciliated zoospores, and of a perfect sexual system like that of *Vaucheria*, have dispelled these views and given the plant what would appear to be its true position. Apropos of the want of colour in this parasitic form, it will be noticed that *Chytridium* and *Pythium*, which are parasitic genera, in the preceding family, are also without colour.

FAMILY XII.—*Zygnemaceæ*. (Plate XII.)

Plants consisting of cylindrical articulated filaments, with the green contents usually disposed in elegant patterns. Reproduction is effected by the phenomenon of conjugation, the whole contents of each pair of united cells being converted into a spore. The particulars of the manner in which this process takes place will be seen in the definitions of the following genera:—

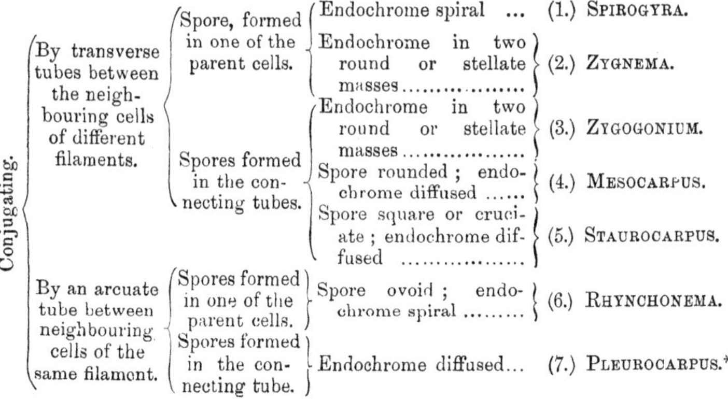

Conjugating.	By transverse tubes between the neighbouring cells of different filaments.	Spore, formed in one of the parent cells.	Endochrome spiral ...	(1.) SPIROGYRA.
			Endochrome in two round or stellate masses..................	(2.) ZYGNEMA.
		Spores formed in the connecting tubes.	Endochrome in two round or stellate masses.................	(3.) ZYGOGONIUM.
			Spore rounded ; endochrome diffused	(4.) MESOCARPUS.
			Spore square or cruciate ; endochrome diffused	(5.) STAUROCARPUS.
	By an arcuate tube between neighbouring cells of the same filament.	Spores formed in one of the parent cells.	Spore ovoid ; endochrome spiral	(6.) RHYNCHONEMA.
		Spores formed in the connecting tube.	Endochrome diffused...	(7.) PLEUROCARPUS.*

1, 2, and 3, *a.* Original filament ; *b.* Conjugation ; 5 and 6, *a.* and *b.* Different species.

* Not figured.

FAMILY XIII.—*Confervaceæ.* (Plate XIII.)

Plants composed of cylindrical cells forming articulated filaments, simple or branched, with a very delicate gelatinous coat. The cell contents are usually green, rarely brown or purple, often assuming peculiar patterns, and ultimately converted into Zoospores, with two or four cilia, from which the filaments are reproduced.

From a fresh-water point of view, only three genera appear to be of importance—viz., *Cladophora, Rhizoclonium,* and *Conferva;* and even these may all be yet included in the succeeding family.

All the species with branching filaments may be referred to the genus *Cladophora ;* for though many species of *Rhizoclonium* have short root-like branches, it so happens that those found in fresh water have simple filaments, which are best distinguished by their decumbent habit from the simple filaments of *Conferva.*

Cladophora glomerata occurs in dark green wavy skeins in pure running water, and

" *crispata* (*b*) in yellowish or dull green strata, is common in fresh, though frequently also in brackish, water.

Rhizoclonium rivulare is found in *fine bright green* bundles, 2–3 feet long, in streams and rivers, &c.

" *implexum* on mountain rocks.

Conferva bombycina. Cells four or five times as long as broad in a *yellowish green* cloudy stratum in stagnant water.

" *floccosa* (*a*). Cells once or twice longer than broad, with circumscissile dehiscence, everywhere common in pools and still water.

Family XIV.—*Œdogoniaceæ.* (Plate XIII.)

Articulated filamentous plants, simple or branched, exhibiting much variety in their means of reproduction. Thus, the whole contents of a cell produce zoospores with a rich growth of cilia, and sporangial cells develop large resting-spores; while antheridial structures are present either on the ordinary filaments or in dwarf parasitic plants. The filaments grow by a rather peculiar process, commencing with circumscissile division of the cellulose coat of the interstitial cells, which thus permits of the growth or extension of the primordial utricle, or under coat, and the formation of a new septum. A cementing band of cellulose repairs the gap between the divided borders of the parent cell, leaving an annular impression to record the fact, and the repetition of the process produces a repetition of the rings, which always characterize even fragments of these plants.

The two genera are easily distinguished, the filaments of *Œdogonium* (*c*) being simple, and those of *Bulbochæte* (*d*) branched and bearing bristle cells with a bulbous base.

The species of *Œdogonium* abound in fresh water under almost all conditions, in lakes, ponds, pools, ditches, streams, and in tanks and cisterns. *Bulbochæte setigera* (*d*), apparently the only reliable species of the genus, grows luxuriantly upon other fresh-water plants.

Family XV.—*Chætophoraceæ.* (Plate XIII.)

These are very beautiful, branched and articulated, filamentous plants, enveloped in gelatinous matter, and made up of cells in single series. Some are free, with a straight central axis; while others are fixed with depressed radiating branches, or forming a discoidal frond. The tapering extremities of the branches in some instances are quite bristle-like, affording one of the distinctive characters of the family. Bristles of an inarticulate kind, however, arise from the articulations in certain genera. Finally, spores and four-ciliated zoospores are formed from the contents of the joints.

Draparnaldia presents a central axis of large colourless cells, with tufts of smaller branches at the articulations. In *Chætophora* (*c*) the filaments are branched and setigerous, indefinitely embedded in gelatinous matter. In *Coleochæte* the frond is discoidal and adherent, composed of radiating dichotomously branched filaments and the bristles springing from the back of the joints are sheathed at the base.

FAMILY XVI.—*Batrachospermaceæ.*

These plants are evidently very closely allied to the *Chæto-phoraceæ,* and the name is derived from the resemblance which their beaded filaments have to frog's spawn. The central axis consists of a single series of cells, with an investment of ad-pressed filaments descending from joints or nodes, occurring at stated intervals, and also giving rise to dense whorls of ex-ceedingly delicate moniliform branches. Some of these latter produce spores at their extremities, whilst others form trans-parent capillary points. The spores form agglomerated masses at the nodes.

In *Batrachospermum* the ramuli are moniliform, while in *Thorea* they are cylindrical. These plants are exclusively aquatic, but chiefly found in pure and gently running water.

FAMILY XVII.—*Characeæ.*

In this interesting family, while the vegetative apparatus is of a very simple type, the generative system is more highly developed than that of any of the preceding forms. These plants consist of a number of large tube-like cells, forming a central axis, and whorls of similar, but smaller cells at the nodal points. So far, this description will answer the genus *Nitella,* which may attain a length of several inches; but in *Chara* an additional envelope is furnished to the central stem by closely applied tubular cells passing from the nodes in both directions, and meeting at the middle of the internodes.

The antheridia and germ cells are here respectively named *globules* and *nucules.* Eight triangular valves radially fluted, and numerous confervoid filaments with antherozoids in the cells, make up the globule; while fine spirally-twisted tubes form the investment of the nucule. So short a notice of these organs is only given to facilitate their recognition when detached.

Notes on the habitat of the Fresh-water Algæ, in relation to the import of their presence in drinking water under examination.

It is important to know that, not only is there a general geographical distribution of aquatic plants in the larger districts of the globe, but also a more restricted localization of certain species, by mere casualty; whilst the special habitat of others is determined by the fulfilment of conditions most favourable to their modes of development and habits of life. Thus, some may be found in running or gently moving water, some in still depths, some in the pool, the pond, the lake; others in the streamlet, the brook, the river; some in regions of death and decay, and others with purer surroundings. There would therefore appear to be good promise in the study of this department of Botany, including not only the recognition of any organisms that may reveal themselves, but the hygienic import of their presence.

The *Palmellaceæ, Coccochloris, Chlorococcum,* &c., appear to depend largely upon the rains, both for their propagation and diffusion, and the moisture surrounding them must be sufficiently persistent to favour the development of their outer gelatinous investment. In the absence of the requisite moisture and consequently of the gelatinous envelope, these humble plants present the appearance of a green efflorescence. How far they may invade Lichens, as casual parasites mistaken for gonidia, is yet an unsettled question.

While the *Desmidiaceæ* are, perhaps with very few exceptions, essentially aquatic, the more striking forms of *Diatomaceæ* are marine. As Desmids seem to love pure water, and usually rest mechanically upon the placid bottom of such spots as are not affected by the constant motion or change going on in their immediate vicinity, their presence might be regarded as a favourable indication.

The *Diatomaceæ* are more widely diffused than any other form of vegetation. They flourish both in standing and running water, and even on the bare ground. In South America some take up their abode amongst lichens upon the trunks of trees. Certain species moreover are found in thermal springs, and others in the pancake ice of the South Pole.

Bacteria are so invariably associated with the decomposition, or rather putrefaction of animal and vegetable matter, that this change is supposed by many to be incapable of taking place without their presence and rapid development.

The *Oscillatoriaceæ* are ubiquitous as a family, though many of them are very definitely distributed; thus, the *Rivulariæ* appear to be confined to Northern regions. They are often found on the stumps of aquatic plants, on rocks in rapid streams, and sometimes where they are exposed to the force of cataracts. They frequently also indicate calcareous water, and crystals of carbonate of lime may be deposited in their substance. A very pretty species presents the appearance of minute green stars upon the surface of lakes. In India *Oscillatorians* ascend to 17,000—18,000 feet above the level of the sea. The *Zygnemaceæ* reach a height of 15,000 feet amongst the Himalayas. They are found in rivers and running waters. The *Confervaceæ* occur in similar situations, but their habitat is in general more varied.

Siphonaceæ (Vaucheria for example) form large tufts in mud whether impregnated with salt or fresh water; they also abound in pools and damp soils. *Codium amphibiorum* affects turf banks at high water, and other members of the family are altogether marine.

Chætophoraceæ and *Batrachospermaceæ* occur in gently moving pure water, of which their presence would therefore afford some indication.

Characeæ inhabit still, rather than moving water. One of the species may be met with in brackish ditches, and the occurrence of these plants can scarcely be regarded as a guarantee of the purity of the water. Many of them however have the property of fixing carbonate of lime, by which they have acquired the name of "stoneworts."

B. *Living Animals.*

The smaller, or microscopic Fauna of the fresh water, as might be expected, is rather comprehensive, including representatives of all the sub-kingdoms and many of the classes of Invertebrata. This will be seen at a glance in the annexed Table, the arrangement of which will be followed in the succeeding pages.

To facilitate the recognition of objects under examination the following definitions of the five sub-kingdoms should be carefully studied, after which it will be found comparatively easy to refer each organism to its proper position in the Table —noting that the definitions apply more particularly to the fresh-water forms.

I. The *Protozoa* (Siebold) are small or minute bodies, either more or less partaking of the character of simple cells, furnished with vibratile cilia or flagella; or resembling merely the contents of cells, destitute of an integument, but capable of throwing out mobile extensions of the sarcode, or gelatinous substance of their bodies. The members of this latter class are named *Rhizopoda,* from the root-like form of the locomotive processes, while those of the former constitute the *Infusoria,* so abundant in organic infusions.

II. The *Cœlenterata* (Frey and Leuckart) are distinguished by having the lining membrane of the stomach continuous with that of the body cavity. They are represented by the single class *Hydrozoa,* including the *Fresh-water Polypes.*

III. The *Annuloida* (Huxley) embrace all the worm-like animals which are not true *Annelida* and the Wheel animalcules. Thus, in the Table the class *Scolecida* (σκώληξ, a worm) includes the ciliated Flatworms (*Turbellaria*), the Threadworms (*Nematoda*), and the Rotifers (*Rotifera*).

IV. The *Annulosa* are distinctly ringed, or segmented animals. They are divided into two classes—viz., those which are without articulated limbs (*Anarthropoda*), including only the *Annelida;* and those which have articulated limbs (*Arthropoda*). The latter are still further divided into the *Crustacea* (with limbs varying in number), the *Arachnida* (or spiders with eight limbs), and the *Insecta* (or insects proper, with six limbs).

V. The *Mollusca,* as the name implies, are soft-bodied animals, usually protected by a testaceous covering or shell. They are divided into an inferior class (*Molluscoida*) taking in the *Limniades* or fresh-water *Polyzoa,* and a superior one (*Mollusca* proper), represented by the fresh-water snails and mussels, or shellfish so-called.

As all the classes have been named and sufficiently characterized in the foregoing definitions, attention may next be directed to the Orders and the illustrative, or more usual genera.

In order to present a bird's-eye view of the subject to the student, the following Table of Classification has been drawn up, with short definitions of the technical terms employed.

A General Table of Classification of Animal Forms, with short Explanatory Notes embodied.

Sub-Kingdom.	*Division.*	*Class.*	*Sub-Class.*	*Section.*	*Order.*	*Genus.*
					(a) Radiolaria (Pseudopodia radiate.)	ACTINOPHRYS.
					(b) Reticularia (Ps. reticulate.)	GROMIA.
I. PROTOZOA (The first, or simplest forms of animal life.)		1. *Rhizopoda* (Like the contents of a cell, throwing out and retracting root-like variable extensions of the body, named pseudopodia.)			(c) Lobosa (Ps. lobed or digitate.)	AMŒBA.
					(d) Spongida (Sponges.)	SPONGILLA.
		2. *Infusoria* (Cell-like, with locomotive whip-like organs, or cilia.)			(a) Flagellata (With whip-like organs.)	PERANEMA.
					(b) Ciliata (With cilia.)	PARAMECIUM.
					(a) Hydrida (Hydra-like.)	HYDRA.
II. CŒLENTERATA (Stomach and body-cavity in communication = Polypes.)		1. *Hydrozoa* (Polypes of the Hydra type.)	A. *Hydroida*		(b) Corynida (Coryne-like.)	CORDYLOPHORA.
			A. *Platyelmia* (Flat worms.)		(a) Turbellaria (Ciliated worms.)	PLANARIA.
III. ANNULOIDA (Like Annulosa.)		1. *Scolecida* (Worms.)	B. *Nematelmia* (Thread-worms.)		(b) Nematoda (Smooth thread-worms.)	ANGUILLULA.
			C. *Rotifera* (Wheel animalcules.)		(c) Rotifera (Wheel animalcules.)	ROTIFER.
					(a) Hirudinea (Leeches.)	GLOSSIPHONIA.
IV ANNULOSA (Ringed or segmented.)	A. ANARTHROPODA (Limbs inarticulate.)	1. *Annelida* (Body ringed.)		*Abranchiata* (Without branchia.)	(b) Oligochæta (Few bristled.)	NAIS.

IV. ANNULOSA ... (Ringed or segmented.) (*Continued.*)	B. ARTHROPODA ...... (With articulated limbs—Articulata.)	1. *Crustacea* .... (Shell coated = crabs, lobsters, and their allies.)	A. *Entomostraca* ..... (Literally, shelled insects.)	1. *Lophyropoda* ...... (Stiff hair-footed.)	(*a*) Ostracoda .... (With bivalve shells.)	CYPRIS.
					(*b*) Copepoda .... (Oar-footed.)	CYCLOPS.
				2. *Branchiopoda* ...... (Gill-footed.)	(*a*) Cladocera .... (Branch-horned.)	DAPHNIA.
					(*b*) Phyllopoda.............. (Leaf-footed.)	BRANCHIPUS.
			B. *Malacostraca* ...... (Soft-shelled, as compared with the testaceous mollusca.)	*Edriophthalmata* (With sessile eyes.)	(*a*) Isopoda .... (Equal footed.)	ASELLUS.
					(*b*) Amphipoda .... (Feet turned different ways.)	GAMMARUS.
		2. *Arachnida* .... (Spiders and their allies.)	A. *Trachearia* .... (Breathing by the general surface, or trachea.)	....	(*a*) Tardigrada.............. (Water bears.)	MACROBIOTUS.
					(*b*) Acarina .... (Water mites.)	HYDRACHNA.
		3. *Insecta* .... (Six-legged articulata = Insects properly so called.)	....	....	(*a*) Coleoptera .... (Beetles.)	DYTISCUS.
					(*b*) Hemiptera .... (Bugs.)	NOTONECTA.
					(*c*) Trichoptera .... (Caddice flies.)	PHRYGANEA.
					(*d*) Neuroptera .... (Dragon flies.)	LIBELLULA.
					(*e*) Diptera .... (Gnats, &c.)	CULEX.
					(*f*) Aptera..................... (Wingless.)	ATAX.
V. MOLLUSCA ... (Soft bodied.)	A. MOLLUSCOIDA ... (Like Mollusca.)	1. *Polyzoa* .... (Numerous by budding, with ciliated tentacula.)	....	....	(*a*) Hippocrepia .... (Horseshoe-shaped.)	CRISTATELLA.
					(*b*) Infundibulata .... (Funnel-shaped.)	PALUDICELLA.
	B. MOLLUSCA .... (Proper) = shell-fish.	1. *Lamellabranchiata* (Bivalve, with two-fold gills.)	....	....	(*a*) Asiphonida .... (Without respiratory siphons.)	UNIO.
					(*b*) Siphonida .... (With respiratory siphons.)	CYCLAS.
		2. *Gasteropoda* .... (Univalve, with a creeping disc.)	....	....	(*a*) Prosobranchiata ...... (Gill in front of the heart.)	PALUDINA.
					(*b*) Pulmonifera .... (With a breathing chamber.)	PLANORBIS.

I. Protozoa.

1. *Rhizopoda.* (Plate XIV.)

Besides the *sponges*, which are represented by the genus *Spongilla* (found in still or slowly running waters, on stones, old workwork, &c.), the Rhizopoda admit of distribution into three groups, easily distinguishable by the characters of the *pseudopodia*, or the motile extensions of the body substance already noticed. In the first group or order (*a*) (*Radiolaria*) they are slender and raylike, persistent, or slowly retractile. In the second (*b*) (*Reticularia*) they are firmly branched, more or less intercommunicating, or reticulate; while in the third (*c*) (*Lobosa*) they are lobose or digitate. These Orders correspond very nearly with those adopted by Dr. Carpenter, F.R.S., and will be better understood on inspecting the following synopsis of the genera. They have the advantage, at least, of being simple, though of course they can only be provisional in the present state of our knowledge of the subject.

(a) *Radiolaria.*

Pseudopodia delicate ray-like simple, besetting the spherical surface.

Body	Naked		(1.) Actinophrys.
	With a covering or shell	With fine spiculæ; free ...	(2.) Acanthocystis.
		Fenestrated; with a pedicle	(3.) Clathrulina.

Habitat:—*Actinophrys digitata* amongst marsh plants; *A. Eichornii* on the surface of infusions, and with *A. discus* (*Trichodiscus*) and the other species, amongst confervæ and aquatic plants. *Acanthocystis* and *Clathrulina* occur in bog-water.

(b) *Reticularia.*

Pseudopodia filiform, reticulate, or finely branched; localized, (Body) globose or ovoid.

Pseudo-podia.	Closely reticulated; sarcode reflected over the shell ...		(1.)	GROMIA.
	Finely branched in a bunch	At one end..	(2.)	PLEUROPHRYS.
		At both ends	(3.)	AMPHITREMA.

Habitat :—*Gromia fluviatilis* on *Ceratophyllum*, *G. hyalina* (with a short neck) in rivulets. *Pleurophrys* and *Amphitrema* in bog-water.

(c) *Lobosa.*

Pseudopodia lobose or digitate, simple or dividing.

Body	With a covering or shell	Pseudopodia, fine and simple, shell flask-shaped	Two or three; subterminal..	(1.)	TRINEMA.
			Numerous; terminal	(2.) & (3.)	EUGLYPHA.
		Pseudopodia, stout and dividing	Shell flask-like .	(4.)	DIFFLUGIA.
			Shell discoidal...	(5.)	ARCELLA.
		Pseudopodium single...	Shell subcubical	(6.)	CYPHIDIUM.
	Naked...	Pseudopodia, variable	....	(7.)	AMŒBA.

Habitat :—*Trinema acinus* and *Euglypha tuberculata* in stagnant water; *Difflugia proteiformis* and *oblonga* amongst *Oscillatoriaceæ*; numerous other species in moist moss at the roots of trees; *Arcella vulgaris* with *Lemnæ* and aquatic plants, *A. aculeata* and *A. dentata* with *Confervæ*; *Cyphidium aureolum* in stagnant water; *Amœba diffluens* on *Lemna* and *A. radiosa* in bog-water.

(d) *Spongida.*

Spongilla, the only fresh-water genus, occurs in little grey or greenish more or less rigid or friable masses, with a spicular framework. They present a superficial or dermal coat, numerous inhalant pores, internal ciliated chambers, and an exhalant aperture. Their grey or green colour is due to the amount of chlorophyll taken into the sarcode or soft substance of the sponge. The silicious spicules which are often present in the sediment of fresh water are—1, birotate; 2, short, crescentic, and echinate; or 3, in the form of stout needles rounded at one end and acute at the other.

2. *Infusoria.*

The heterogeneous members of this class have been of late years divided into three Orders—viz., 1st, the *Ciliata;* 2nd, the *Suctoria;* and 3rd, the *Flagellata.* The *Suctoria,* however, being but phases of *Vorticellina,* a family of *Ciliata,* can scarcely be regarded as a separate Order, and may therefore be rejected. This will leave the *Ciliata* and *Flagellata;* and it may even be said with truth that most of the latter actually belong to the Botanist. Nevertheless, as the object of this work is to facilitate the recognition of the forms themselves, the knotty question of their natural classification must be deferred to a future period, when their whole history will have become better known.

With some few alterations and the exclusion of the foregoing *Rhizopoda,* the following Tables are in accordance with Dujardin's arrangement of the *Infusoria.*

(a) *Flagellata.*

Furnished with one or more flagelliform (whip-like) filaments for locomotion, rarely also with a linear series of cilia.

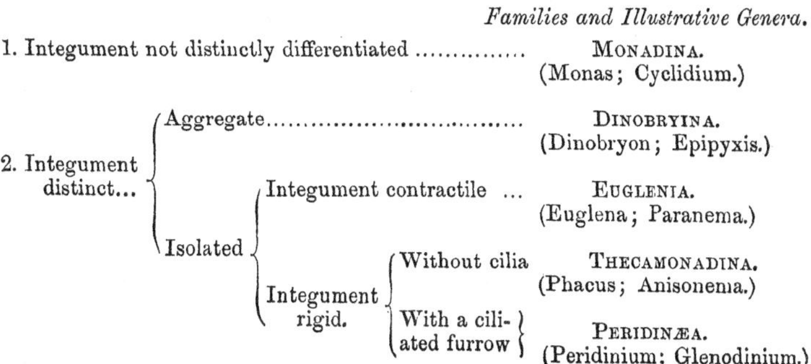

Families and Illustrative Genera.

1. Integument not distinctly differentiated .... MONADINA. (Monas; Cyclidium.)
2. Integument distinct...
   - Aggregate .... DINOBRYINA. (Dinobryon; Epipyxis.)
   - Isolated
     - Integument contractile ... EUGLENIA. (Euglena; Paranema.)
     - Integument rigid.
       - Without cilia THECAMONADINA. (Phacus; Anisonema.)
       - With a ciliated furrow PERIDINÆA. (Peridinium; Glenodinium.)

Habitat :—The *Monadina* are usually found in animal and vegetable infusions, in decomposing water, and especially amongst decaying fresh-water Algæ. *Euglena viridis* abounds in pools, and like *Phacus*, which also affects stagnant water, often imparts its green tint to the surrounding medium. Though the *Peridinæa* may occur in stagnant ponds, they are not to be found in decomposing water or infusions.

(b) *Ciliata.*

Furnished with vibratile cilia, variously distributed, either as connected with the mouth, or the general surface.

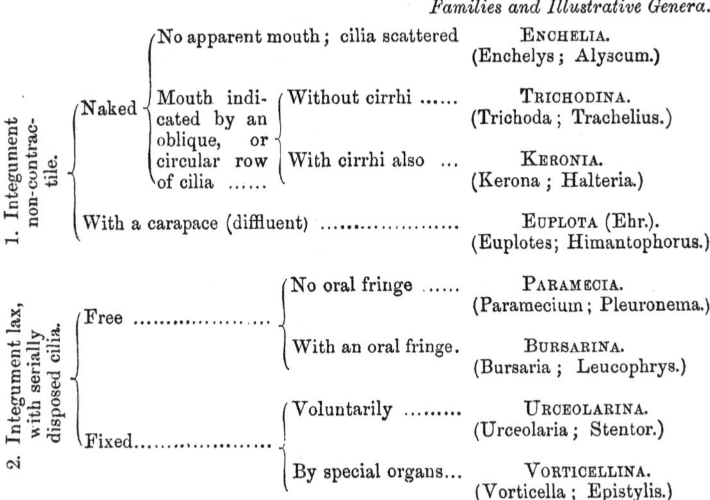

				Families and Illustrative Genera.
1. Integument non-contractile.	Naked	No apparent mouth; cilia scattered		ENCHELIA. (Enchelys; Alyscum.)
		Mouth indicated by an oblique, or circular row of cilia	Without cirrhi	TRICHODINA. (Trichoda; Trachelius.)
			With cirrhi also ...	KERONIA. (Kerona; Halteria.)
	With a carapace (diffluent)			EUPLOTA (Ehr.). (Euplotes; Himantophorus.)
2. Integument lax, with serially disposed cilia.	Free		No oral fringe	PARAMECIA. (Paramecium; Pleuronema.)
			With an oral fringe.	BURSARINA. (Bursaria; Leucophrys.)
	Fixed....................		Voluntarily	URCEOLARINA. (Urceolaria; Stentor.)
			By special organs...	VORTICELLINA. (Vorticella; Epistylis.)

Habitat :—The notable presence of the *Ciliata* would indicate not only stagnant water, but such as may contain organic matter in solution to some relative extent, not yet precisely determinable. Some *Paramecia*, however, as *Amphileptus*, are found in clear marsh water and streams running between aquatic plants. Some *Enchelia* and *Keronia* occur in water with decomposing vegetable matter; and the *Vorticellina* also abound in vegetable infusions, though several are parasitic on *Entomostraca* in comparatively good water.

Some *Bursarina* present themselves in the intestine of the Frog, and of *Nais;* and *Urceolaria pediculus* may be seen gliding over the ciliated surface of *Planaria* (see also the note appended to the Table of the species of *Hydra* below).

The following Tables of classification will form the most convenient description of the figures.

Flagellata. *Monadina.* (PLATE XIV.)

Isolated.	A single flagellum ...	Arising in front	Mobile throughout...	1. MONAS.
			Mobile at the end ...	2. CYCLIDIUM.
		Arising just behind the beak ...		3. CHILOMONAS.
	A second filament ...	Arising laterally		4. AMPHIMONAS.
		Posteriorly		5. CERCOMONAS.
		In front, but trailing		6. HETERAMITA.
	Two equal filaments at the curved angles in front.........			7. TREPOMONAS.
	Four equal filaments in front, and two thicker ones posteriorly			8. HEXAMITA.
Aggregate.	Group originally fixed on a branched axis			9. ANTHOPHYSA.
	Group always free, revolving			10. UVELLA.

Dinobryina.

With an urceolate carapace.	Single, without an eye-speck, and commonly without a flagellum	1. EPIPYXIS.
	Forming a branched aggregate, with both eye-speck and flagellum	2. DINOBRYON.

Euglenia.

Attached				1. COLACIUM.
Free......	No visible flagellum; two eye-spots			2. DISTIGMA.
	One flagellum	With an eye	With a tail-like process	3. EUGLENA.
			Without a tail	4. AMBLYOPHIS.
		Without an eye		5. PERANEMA.
	Two flagella		Green or red; an eye-spot	6. CHLOROGONIUM.
			Colourless; no eye-spot	7. ZYGOSELMIS.
	Several flagella........			8. POLYSELMIS.

Thecamonadina. (Plate XV.)

Flagellum	Single	Body ovoid or globular	Integument hard ...	1. Trachelomonas.
			Integ. membranous	2. Cryptomonas.
		Body depressed or foliaceous	With a tail	3. Phacus.
			Without a tail	4. Crumenula.
	Two, one trailing			5. Anisonema.

Peridinæa.

Carapace...	Setaceous	With an eye-spot ...	1. Chætoglena.
		No eye-spot	2. Chætotyphla.
	With a ciliated furrow ...	With an eye-spot ...	3. Glenodinium.
		No eye-spot	4. Peridinium.

Ciliata.—Enchelia. (Plate XV.)

Body..	Partially ciliated ...	Cilia at one end	1. Acomia.
		Cilia in a longitudinal furrow	2. Gastrochæta.
	Ciliated all over	Cilia all alike	3. Enchelys.
		With a trailing filament also	4. Alyscum.

Trichodina.

Cilia	Covering the body	Forepart curved like a hatchet	1. Pelecida.
		Neck long and slender	2. Dileptus.
		Upper lip prolonged into a proboscis ...	3. Trachelius.
	In front, chiefly	On one or both sides directed forwards ...	4. Acineria.
		In one row, directed backwards	5. Trichoda.

Keronia.

Having styles		3. Urostyla.	5. Stylonychia.
,, cirrhi	1. Halteria.		
,, cilia			
,, setæ	2. Oxytricha.	4. Kerona.	
,, uncini			

Euplota.

Body depressed	Having hooks but no styles	1. Himantophorus.
	With both styles and hooks	2. Euplotes.

Paramecia. (Plate XVI.)

Teeth present.	Mouth lateral	Body	Lamelliform, frontal margin produced on one side		1. Chilodon.
			Globular		2. Nassula.
	Mouth terminal	Body	Globular		3. Prorodon.
Without teeth.	Mouth lateral	With appendages	With vibrating lips		4. Glaucoma.
			Lip inferior, projecting		5. Colpoda.
		Without appendages	Mouth within a longitudinal oblique fold		6. Paramecium.
			Mouth impinging on the margin		7. Panophrys.
	Mouth terminal	Body	Globular by contraction		8. Holophrya.
			Flask-shaped.	Pointed behind.........	9. Trachelocera.
				Rounded behind	10. Lacrymaria.

Bursarina.

Body	Short purse-like or moderately elongated......	Acuminate posteriorly, with a frontal eye-spot..........................	1. Ophryoglena.
		Rounded posteriorly, no eye-spot; a spiral row of cilia in front, ending in a large mouth	2. Bursaria, and 3. Leucophrys (Ehr.)
	Much elongated, cylindrical ...	Mouth in advance of the middle, at the end of a row of cilia	4. Spirostomum.

Urceolarina.

Clustered vorticella-like animals in gelatinous masses			1. Ophrydium.
Solitary	With a crown of cilia at both ends	Body short and discoidal	2. Urceolaria.
	Crown of cilia in front only ...	Body trumpet-shaped, ciliated all over, mouth spiral	3. Stentor.
		Body bell-shaped, smooth; tail subulate	4. Urocentrum.

Vorticellina.

Stalk present.	Bodies all uniform ...	Stalk spirally flexible	Simple ...	1. VORTICELLA.
			Branched	2. CARCHESIUM.
		Stalk inflexible		3. EPISTYLIS.
	Bodies of two shapes	Stalk inflexible		4. OPERCULARIA.
		Stalk spirally flexible...............		5. ZOOTHAMNIUM.

Symmetrical Forms. (PLATE XVII.)

The genera (1) Ichthydium, (2) Chætonotus, (3) Coleps, and (4) Planariola are placed by Dujardin as an appendix to the Ciliata, though they have no natural affinity *inter se,* on account of exhibiting a bilateral symmetry, which singularly enough, is wanting in all the other ciliated Infusoria.

II. CŒLENTERATA. (PLATE XVII.)

The only *Cœlenterata* occurring in fresh water are members of the sub-class *Hydroida,* the two first orders of which—viz., *Hydrida* and *Corynida,* are represented by the respective genera *Hydra* and *Cordylophora.*

(a) *Hydrida.*

The first Order is distinguished by the Polypites or separate Zooids being single and locomotive, with a sucker disc at one end, and an oral orifice at the other, surrounded with tentacula. The integument never developes a sclerous layer, and the reproductive organs appear as simple external processes of the body.

Table of the species of the genus *Hydra*.

Body			
Cylindrical or insensibly narrowed towards the base.	Tentacula shorter than the body, smaller at the base................	(1.)	*H. viridis.* (Leaf green.)
	Tentacula as long as or longer than the body, tapering to the end ...	(2.)	*H. vulgaris.* (Yellowish or red.)
Attenuated below in a marked degree.	Tentacula longer than the body ...		*H. attenuata.* (Pale olive green.)
	Tentacula several times longer than the body		*H. fusca.* (Brown or greenish.)

Habitat:—In ponds and still waters on *Lemna* and aquatic plants.

Note.—Parasitic Infusoria are often found upon these Polypes—viz., *Kerona polyporum* or *H. vulgaris* and *H. fusca*; and *Urceolaria pediculus* or *H. vulgaris* and *H. viridis*. Their presence, however, would indicate impurity of the water and an unhealthy condition of the Polypes themselves.

(b) *Corynida.*

In this, the second order, the Polypites are either single or two or more connected by a common substance or "*Cœnosarc*," always fixed at the base, and usually developing a firm outer layer or "*Polypary*." The reproductive organs or "*Gonophores*" arise either from the Polypites, the Cœnosarc, or the so-called "*Gonoblastidia*."

Genus *Cordylophora* (Allman.)

Polypary horny, branched, and rooted by a creeping tubular stolon; polypes ovoid, with a small mouth, and scattered filiform tentacula.

(3.) *Cordylophora lacustris* was the only species known to Allman, but lately a second, *C. rivularis*, has been added.

III. ANNULOIDA.

1. *Scolecida.*

(a) *Turbellaria.* (PLATE XVII.)

Non-parasitic ciliated worms. Some of these are bisexual, with a single alimentary or oral opening, and constitute the first sub-order (*Planarida*), including fresh-water species, whilst others are unisexual, with two alimentary openings, and form a second sub-order (*Nemertida*) altogether marine.

Planarida.

				Illustrative Genera.
Intestine	Straight—*Rhabdocœla*	Concatenated		(1.) DEROSTOMUM.
		Single	Mouth near the fore part	(2.) PROSTOMUM.
			Mouth near the middle	(3.) MESOSTOMUM.
	Ramose—*Dendrocœla*			(4.) PLANARIA.

Habitat:—All in ponds and gently moving deep water amongst aquatic plants.

(b) *Nematoda.* (PLATE XVIII.)

The non-parasitic threadworms composing the family of *Anguillulidæ* are very frequently met with in fresh waters. The vinegar eel (*Anguillula aceti*), and sour paste eel (*A. glutinis*), and the *Tylenchus* (or so-called *vibrio*) *tritici*, invading the ears of corn, belong to this family. *Anguillula fluviatilis* is colourless or white, about fifteen times as long as it is broad, with a fusiform œsophagus, expanding posteriorly into a much larger stomach. 1, Anguillula found in bilge-water; 2, A. aceti; 3, A. fluviatilis.

The *Anguillæ* are readily confounded with the *Enoplidæ,* a family of minute parasitic *Nematodes,* infesting the intestine of aquatic larvæ and other small animals, but often found free in the water.

(c) *Rotifera.* (PLATE XVIII.)

The Wheel Animalcules, so called on account of the deceptive appearance produced by the regular and consecutive action of the vibratile cilia fringing the head-lobes. These latter may be simple, sinuated, lobed or divided, and are capable of retraction and protrusion. The alimentary system is usually distinct, with a dental apparatus and two orifices, and the sexes are separate.

As a whole these little creatures present superficial points of resemblance to the *Entomostraca,* to which the character of their segmentation makes a nearer approach than that of any *Annelida.* Indeed, they have been rather appropriately named *Cilio-crustaceans* by Leydig. Dujardin grouped them in the following simple manner:—

	Illustrative Genera.
1. Those that are fixed ...	*Floscularia. Melicerta.*
2. Those that swim only ...	*Brachionus. Furcularia. Albertia.*
3. Those that both swim and crawl	*Rotifer.*

Ehrenberg's arrangement, though perhaps more artificial, may still be found more convenient for the recognition of genera.

Rotifera.

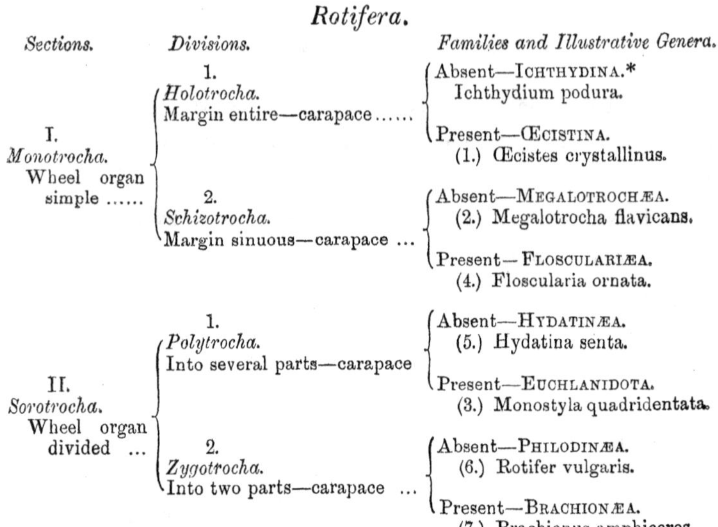

Sections.	*Divisions.*	*Families and Illustrative Genera.*
I. *Monotrocha.* Wheel organ simple	1. *Holotrocha.* Margin entire—carapace......	Absent—ICHTHYDINA.* Ichthydium podura.
		Present—ŒCISTINA. (1.) Œcistes crystallinus.
	2. *Schizotrocha.* Margin sinuous—carapace ...	Absent—MEGALOTROCHÆA. (2.) Megalotrocha flavicans.
		Present—FLOSCULARIÆA. (4.) Floscularia ornata.
II. *Sorotrocha.* Wheel organ divided ...	1. *Polytrocha.* Into several parts—carapace	Absent—HYDATINÆA. (5.) Hydatina senta.
		Present—EUCHLANIDOTA. (3.) Monostyla quadridentata.
	2. *Zygotrocha.* Into two parts—carapace ...	Absent—PHILODINÆA. (6.) Rotifer vulgaris.
		Present—BRACHIONÆA. (7.) Brachionus amphiceros.

* Ichthydium podura and Chætonotus larus will be found amongst the symmetrical Infusoria (Plate XVII., 1 and 2), to which Dujardin has referred them. Their true position, however, has scarcely yet been determined.

IV. ANNULOSA.

A. *Anarthropoda.* 1. *Annelida.* (PLATE XVIII.)

(a) *Hirudinea.*

All the Leeches have a more or less sucker-like mouth, and are also furnished with a disk-shaped caudal sucker; and although the body is finely annulated, it is divided into larger somites or segments like other annelida. The nervous system is highly developed, and the sexes are combined in the same individual; but neither self-impregnation nor reproduction by fission or gemmation has been observed in any case. The fresh-water types may be thus arranged:—

			Illustrative Genera.
Eyes	10 in number	With jaws and teeth, but no proboscis........................	... HIRUDO.
	Less than 10	With no teeth or proboscis ...	(2.) NEPHELIS and BDELLIA.
		With proboscis, but no teeth .	(3.) GLOSSIPHONIA.

Habitat:—In ponds and lakes and slowly-moving waters.

(b) *Oligochæta.*

The *Oligochæta,* or *Setigera,* include the *Earthworms* (*Lumbricini*) and the true water worms (*Naididæ*). Their bodies are usually much elongated, and furnished with locomotive chitinous setæ or bristles attached in rows to the sides and ventral surface laterally. The *Lumbricini* are hermaphrodite, and the *Naididæ* unisexual, but the latter also multiply in a remarkable way by gemmation and fission.

		Illustrative Genera.
1. *Lumbricini.* (Terrestrial and aquatic.)	Having four rows of setæ, two dorsal and two ventral, on each side	... TUBIFEX.
2. *Naididæ.* (All aquatic.)	Two rows of setæ, one dorsal and one ventral, on each side; the four first segments without dorsal setæ	(1.) NAIS.
	With ventral setæ only	... CHÆTOGASTER.

In Lamarck's genus *Stylaria* the setæ are very long, and the cephalic segment is produced into a kind of proboscis. The genus *Proto,* founded by Oken, is distinguished by the presence of ciliated tentaculiform processes surrounding the dorsal and subterminal vent, as in Fig. 1 a.

Habitat :—All these little worms live amongst aquatic plants, burrow in the mud, or manufacture little tubes into which they retreat for protection. The *setæ,* but more especially the ventral uncini (1 b), which are usually bifid at the extremity, are frequently found in the sediment of water in which algæ have been kept for some little time.

Note.—In some instances two speck-like eyes are present, and they may be confounded with the aquatic larvæ of insects. They differ, however, in having the setæ implanted beneath the general surface, and the absence of the fine dark ramifications of the trachææ and of oral or cephalic organs of any kind, except the above-mentioned eye-specks.

B. *Arthropoda.*

1. *Crustacea.*

A. *Entomostraca.* (PLATE XIX.)

The first four out of the six orders of Crustacea bearing aquatic genera belong to the sub-class *Entomostraca,* which may be said to consist of an empirical assemblage of usually very small or minute crustaceans, having either less than seven, or more than ten pairs of legs. To this it must be added, that the branchiæ are either attached to the oral organs, constituting the first section *Lophyropoda,* or to the legs, composing the second section *Branchiopoda.* Each of these is still further divided (as in the general Table) into two orders.

1. *Lophyropoda.*

(a) *Ostracoda.*

Body completely enclosed in a bivalve carapace or shell. Legs, 2 or 3 pairs.

	Families.		*Illustrative Genera.*
A single eye.	Cypridæ (Legs, 2 pairs).	Both pairs of antennæ with a tuft of hairs	(1.) CYPRIS.
		Inferior antennæ without the tuft	(2.) CANDONA.
	Cytheridæ (Legs, 3 pairs).	Superior antennæ without the tuft	(3.) CYTHERE.

Habitat :—In ponds and lakes.

(b) *Copepoda.*

Shell jointed, forming a buckler enclosing the head and thorax. Legs, 5 pairs.

		Families.			*Illustrative Genera.*
A single eye.	Both superior antennæ in the male with a swollen joint...	*Cyclopidæ* ...	Foot-jaws, 2 pairs...	Large and branched; ovaries 2	(1.) CYCLOPS.
				Small and simple; ovary 1	(2.) CANTHOCAMPTUS.
	Male with a swollen hinge on right superior antenna only ...	*Diaptomidæ*	Foot-jaws, 3 pairs...	Ovary 1	(3.) DIAPTOMUS.

Habitat :—In ponds and ditches.

2. *Branchiopoda.*

(a) *Phyllopoda.*

Legs from 11 to 60 pairs; joints foliaceous, branchiform.

		Families.		Illustrative Genera.
Body	Naked ...	*Branchiopoda*	Tail simply bifid	ARTEMIA.
			Tail in two distinct pieces	(1.) BRANCHIPUS.
	In a shell	*Aspidephora* ...	....	(2.) APUS.

Habitat :—Respectively in saltpans, ditches, and pools.

(b) *Cladocera.*

Body included in a pseudo-bivalve carapace. Legs, 5 or 6 pairs.

		Families.		Illustrative Genera.	Branches of the inferior antennæ.
A single eye.	Intestine simple, no black spot in front of the eye ...	*Daphnidæ*	Legs, 6 pairs	(7.) DAPHNELLA	2 & 2 jointed
				(8.) SIDA	3 & 2 jointed
			Legs, 5 pairs	(5.) DAPHNIA	4 & 3 jointed
				(6.) BOSMINA	
	Intestine convoluted, a black spot in front of the eye ...	*Lynceidæ*	Legs, 5 pairs	(1.) CHYDORUS	3 & 3 jointed
				(2.) CAMPTOCERCUS .	
				(3.) ALONA	
				(4.) PLEUROXUS	

Habitat :—In ponds, ditches, tanks, and reservoirs; usually in good water.

B. *Malacostraca.* (PLATE XX.)

a. *Edriophthalmata.*

(a) *Isopoda.*

(1) *Asellus aquaticus* appears to be the only fresh-water Isopod. Its distinguishing features are the following:—Superior antennæ, at least as long as the peduncle of the inferior ones. The seven pairs of legs of the order, with the terminal hooks entire; and two bifid needle-like processes at the posterior extremity of the body.

Habitat :—Plentiful in stagnant pools, passing the winter in the mud, from whence it emerges in the spring.

(b) *Amphipoda.*

(1) *Gammarus* is the only genus of *Amphipoda* occurring in fresh water. A short branch arises from the tip of the third joint of the superior antennæ, and the four anterior legs are in the form of small claws with the moveable tip folding on the inner side.

(2) *Gammarus pulex* is the type of the genus, and abundant in fresh-water brooks where there is an accumulation of vegetable débris.

G. fluviatilis, another fresh-water species, is at once distinguished by the presence of a dorsal spine at the posterior border of each abdominal segment.

Note.—In concluding the notice of the Crustacea it must be mentioned that the larvæ of some of the *Oniscidæ* or wood-lice are aquatic.

2. *Arachnida.* (PLATE XX.)

(a) *Tardigrada.*

The water bears are distinguished by having the head marked off from the thorax, while the thorax and the abdomen are confluent. The body is faintly divided into four segments, carrying each a pair of obscurely three-jointed legs, with three or four claws at their extremity. They form but one family, including three genera as under :—

			Illustrative Genera.
Head	With appendages ...	Mouth conical, without sucker or appendages	(1.) EMYDIUM.
		Mouth sucker-like, with palpiform appendages	(2.) MILNESIUM.
	Without appendages	Mouth sucker-like, without appendages	(3.) MACROBIOTUS.

Habitat :—Stagnant water amongst water plants, in wet moss, and even in the gutters of houses, from whence they may be washed into cisterns and waterbutts.

(b) *Acarina.*

In this Order we find the *Hydrachnea* or water mites, with the head, thorax, and abdomen all fused together; the Palpi with the last joint unguiculate or spinous; the eyes two or four, and the legs generally ciliated and natatory, the posterior pair the longest. Of the several genera *Hydrachna* would appear to be the most commonly met with. (1) *Hydrachna globula* is subovate in form, of a rich deep red colour, with two pairs of eyes at a moderate distance apart, and the skin is covered with minute puncta. The generic name *Achlysia* has been given to the hexapod (six-legged) young of this genus, the Nymphs of which are parasitic on aquatic insects. (2) *Hydrachna geographica.* (3) A still more globular form. (4) *Limnochares holosericus*, crawling, not natatory.

Habitat:—In ponds and permanent lodgments of water. *H. globula* uses its legs with great activity, as though running through the water, instead of swimming.

3. *Insecta.*

The more usual aquatic larvæ are of the following Orders, as given by Kirby and Spence, and are sufficiently numerous to suggest that they would be more readily determined by the use of figures than by description, however elaborate.

Families.	*Genera.*
(*a*) *Coleoptera*	DYTISCUS, HYDROPHILUS, GYRINUS, LIMNIUS, PARNUS, HETEROCERCUS, ELOPHORUS, HYDRÆNA.
(*b*) *Hemiptera*	GERRIS, VELIA, HYDROMETRA, NOTONECTA, SIGARA, NEPA, RANATRA, NAUCORIS.
(*c*) *Lepidoptera*	A few (as NYMPHULA).
(*d*) *Trichoptera*	The majority (PHRYGANEA, &c.)
(*e*) *Neuroptera*	LIBELLULA, ÆSHNA, AGRION, SIALIS, EPHEMERA.
(*f*) *Diptera* ...	CULEX and TIPULARIÆ.
(*g*) *Aptera*......	ATAX and some PODURÆ.

The smaller species of water beetles, *Hydrophilus*, *Elophorus*, *Hydræna*, *Parnus*, *Limnius*, and also *Nepa*, walk upon the water. The swimmers generally have the posterior legs fitted for the purpose. Thus, in *Dytiscus* and *Notonecta* they are furnished with a dense fringe of hairs on the shank and foot, and in *Gyrinus* the terminal joints are very much dilated.

Some insects walk and swim upon the surface without diving, as *Gerris lacustris*, the water-bug, which can walk, run, jump, or swim upon the surface.

Hydrometra stagnorum, very slender in form with prominent hemispherical eyes, apparently in the middle of the body, though really on the head, ramble over stagnant water, and *Velia rivulorum* courses rapidly over running streams and rivers.

V. Mollusca.

A. *Molluscoida.*

The *Limniades* or fresh-water Polyzoa are thus characterized. *Polyzoarium* fleshy, spongy, or coriaceous; apertures angular or round, closing when the zooids recede. Tentacula ciliated in a single series, fringing a more or less crescentic lophophore (*Phylactolæmata*), or an orbicular one (*Gymnolæmata*), in both cases including the mouth. The genera *Cristatella* and *Plumatella* are examples of the former group, while *Paludicella* and *Urnatella* represent the latter.

The *Polyzoarium* in the *Cristatellidæ* is membranous, sacciform and free, or floating, while that of the *Plumatellidæ* is fixed, fistular, and confervoid.

Habitat :—Ponds and lakes.

B. *Mollusca* (proper).

The simple recognition of the shell, univalve or bivalve, will suffice for the *Mollusca* proper, or the fresh-water shellfish, so called; conchological works may be consulted if necessary. The following genera occurring either in this or other countries are merely cited as examples.

1. *Lamellibranchiata* (Bivalves).

(a) *Asiphonida, Anodon, Unio,* (b) *Siphonida, Cyclas, Pisidium, Cyrena.*

2. *Gasteropoda* (Univalves).

(a) *Prosobranchiata* (Operculate).
Neritina, Navicella, Paludina, Ampullaria, Hydrobia, Valvata, Melania.

(b) *Pulmonifera* (Inoperculate).
Limnæa, Physa, Planorbis, Ancylus.

MINERAL MATTER.

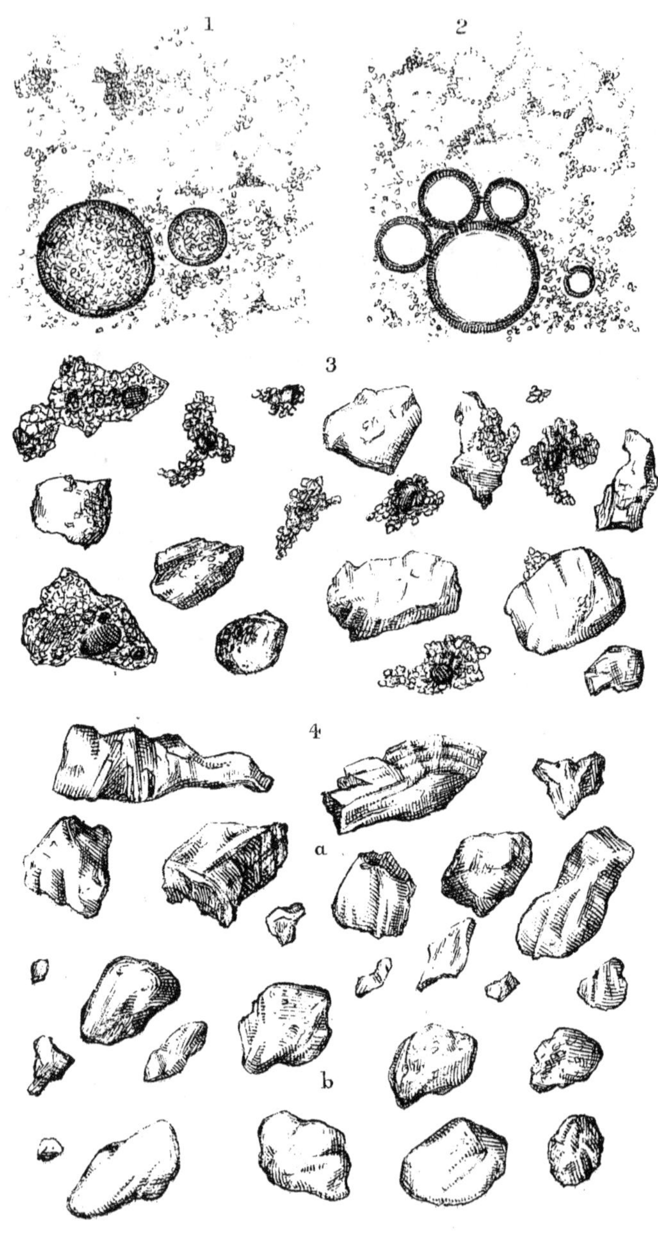

PLATE I.

Mineral Matter.

1. Carbonate of lime, finely divided with vesicles of atmospheric air, between the glass slip and cover.

2. Also carbonate of lime, but with the evolution of carbonic acid gas by the addition of an acid.

3. Fine green mineral particles, cohering as a microscopic breccia, or conglomerate, are here and there mingled with larger and probably more recent sandy granules, preserving their angularity and roughness from fracture; taken from the débris of a well-sinking, at the Royal Victoria Hospital, Netley.

4. Silicious or flinty granules taken from road-side streamlets, (*a*) more recent, and (*b*) of earlier date, having been rounded off and smoothed by rolling and attrition, like microscopic boulders.

VEGETABLE PRODUCTS.

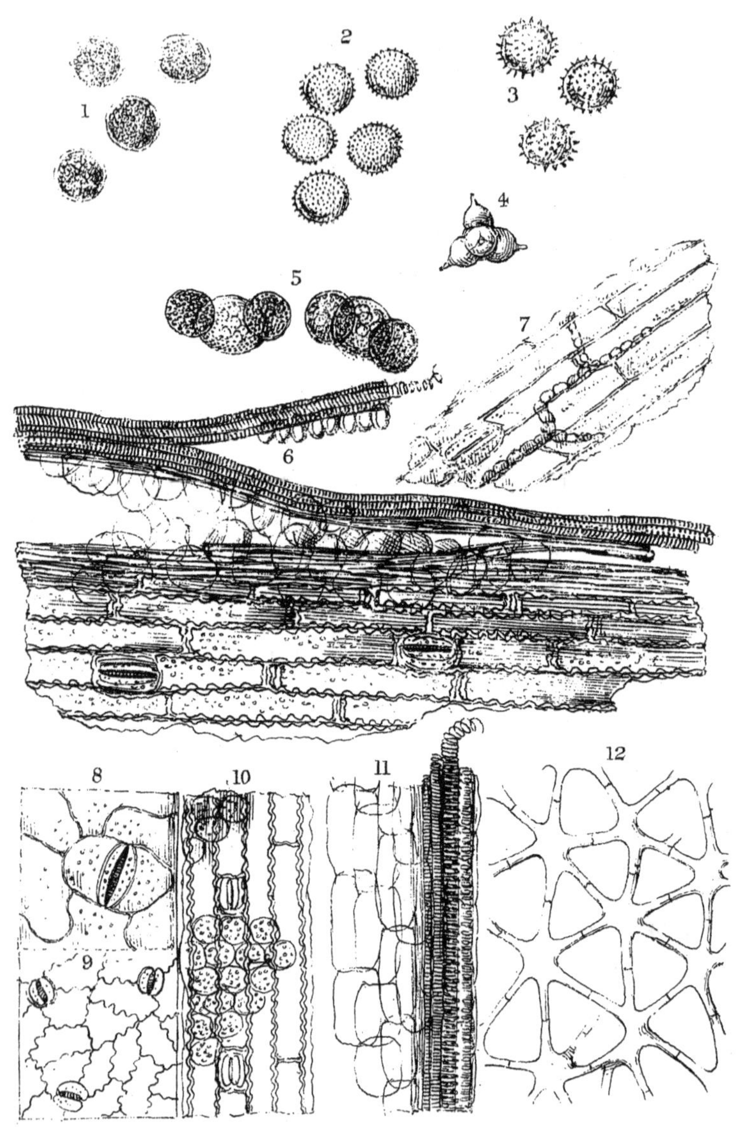

PLATE II.

Vegetable Products.

1. Pollen of Grass.

2. „ of Water-lily.

3. „ of Water-plantain.

4. „ of Rush.

5. „ of Pine.

6. Epidermis, parenchyma, and fibro-vascular tissue of straw.

7. Cuticle of Grass,* with the mycelium of Odium monilioides.

8. Epidermis or cuticle of Water-plantain.

9. Ditto of the lesser Duck-weed.

10. Cuticle of Carex, with stomata, and some of the round subjacent parenchyma cells seen through it.

11. Section of the stem of Carex, showing the large pith or medullary cells, and a bundle of pitted tissue and spiral vessels.

12. Stellate tissue of the pith of the Rush.

---

* The epidermis and other tissues of grasses, as of hay and straw, derived from stable manure which is being constantly dried and powdered on every road, and widely dispersed by the wind, are very frequently present in water to which they may find access.

VEGETABLE PRODUCTS.

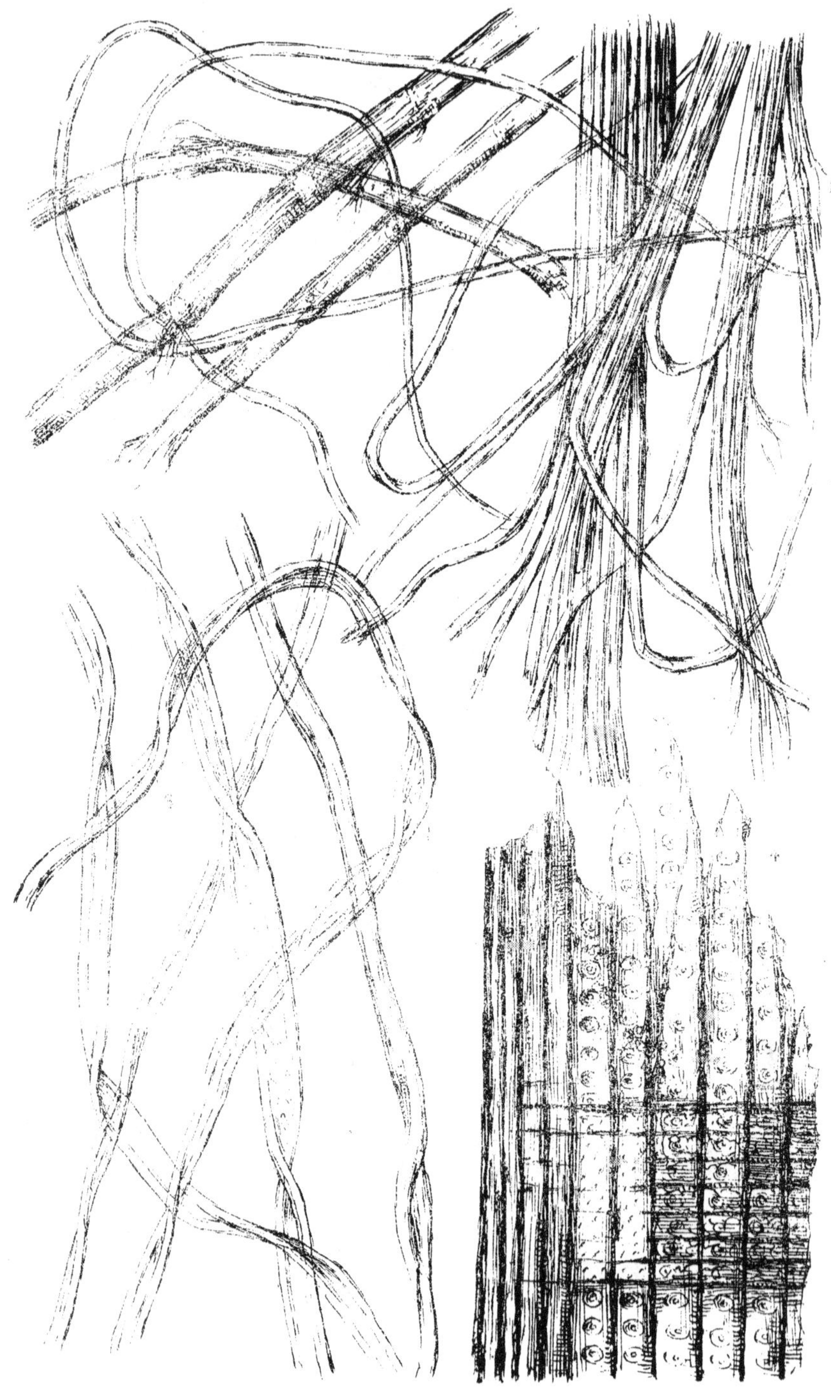

PLATE III.

Vegetable Products indicative of Contamination with House Refuse.

1. Linen fibre.

2. Hemp.

3. Cotton.

5. Chip of deal or pine, with the so-called discoidal tissue, and the silver grain of carpenters passing at right angles to the woody fibres.

ANIMAL PRODUCTS.

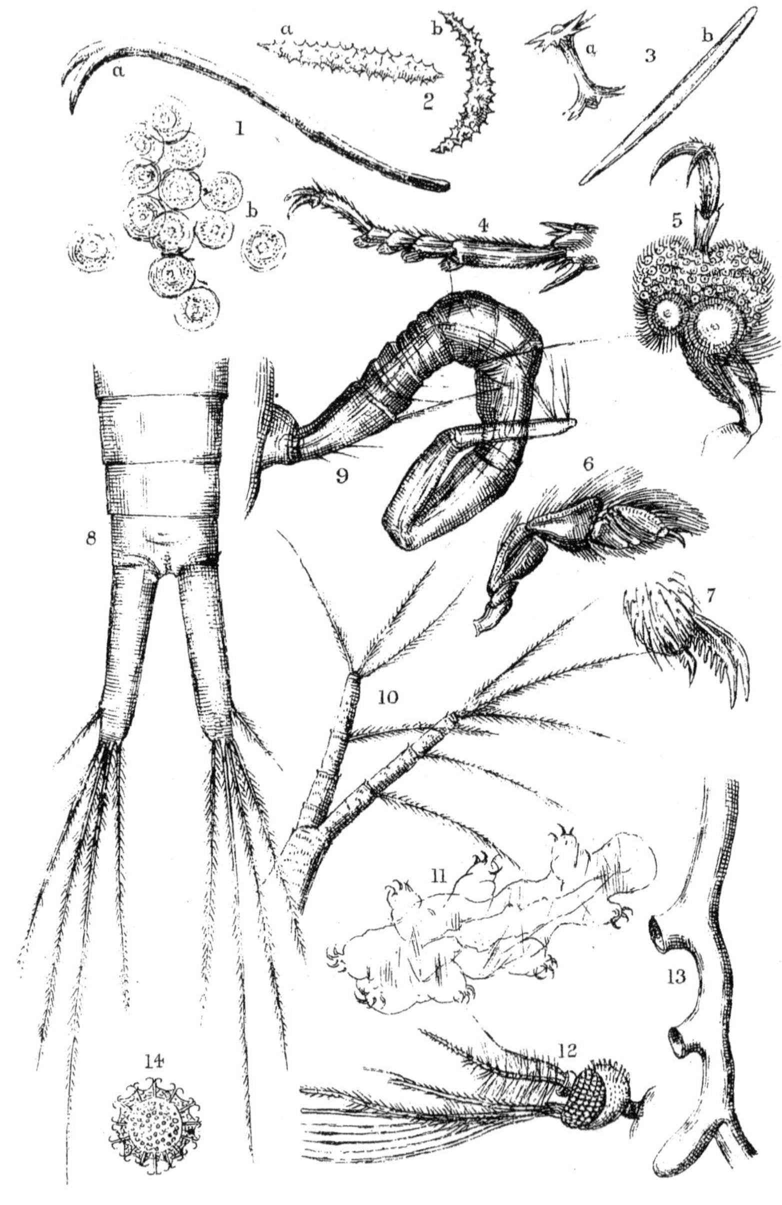

PLATE IV.

Animal Products.

1. (*a*) Ventral hooklet of Nais (a fresh-water annelid); (*b*) liberated ova of the same, often visible when the body of the parent has broken down so as to be indistinguishable.

2. Spiny spicula of Spongilla lacustris, (*a*) straight; (*b*) curved.

3. Spicula of Spongilla fluviatilis, (*a*) birotulate;* (*b*) simple.

4. Part of the leg of a Cockroach.

5. Fore leg of Male Dytiscus.

6. Hind leg of Gyrinus natator.

7. Foot of a Spider.

8. Tail of Cyclops quadricornis (male).

9. Right superior antenna of the same.

10. Inferior antenna of Daphnia pulex.

11. Cast skin of Macrobiotus (Tardigrada).

12. Head and trophi of Gnat (Culex).

13. Portion of the Polypidum of Plumatella (Polyzoa).

14. Egg of Cristatella Mucedo.

---

* The corresponding spicules of the Bombay Tank Sponge, Spongilla Meyeni, form very good objects for the microscope.

ANIMAL PRODUCTS.

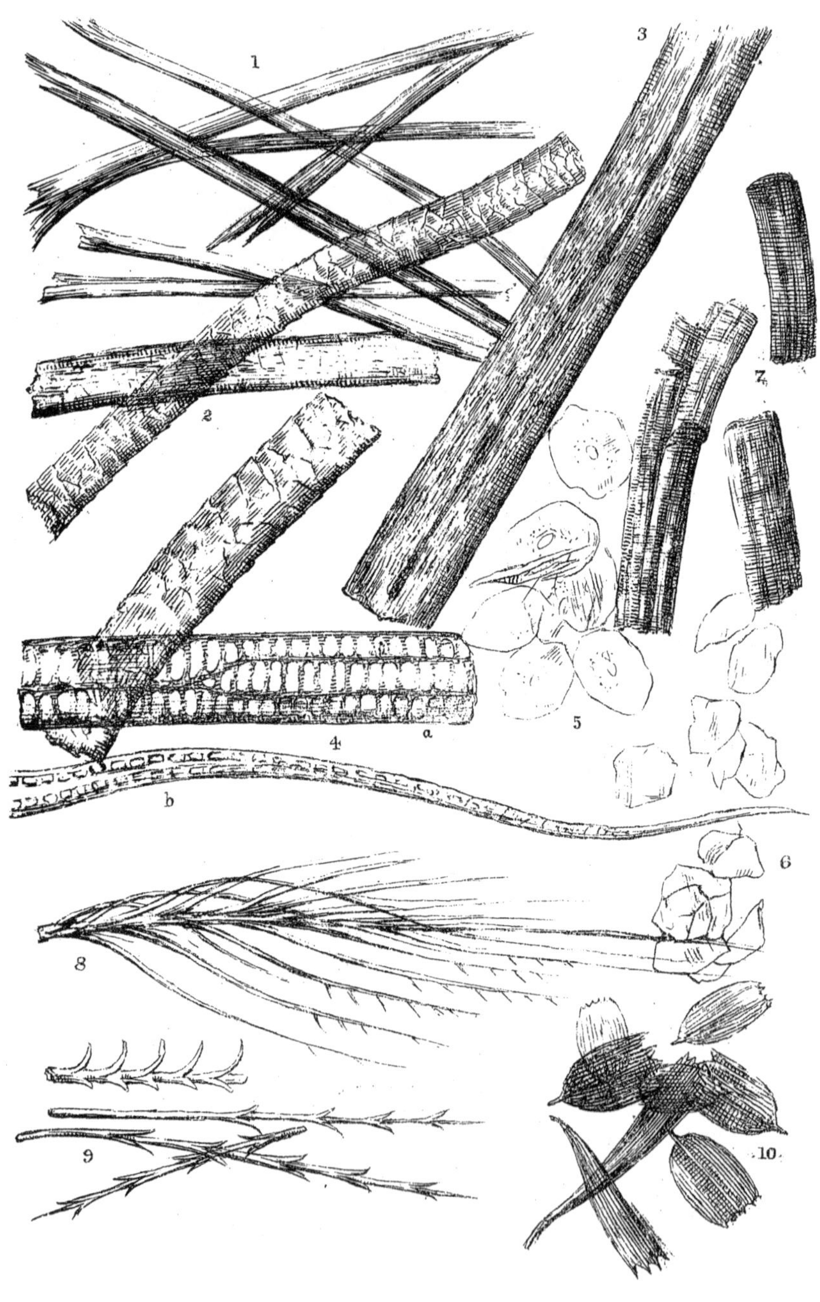

PLATE V.

Animal Products foreign to the Fresh Water.

1. Fibres of silk. 2. Woollen fibre. 3. Human hair.

4. Rabbit's hair, (*a*) the shaft; (*b*) the apex.

5. Nucleated scale-like epithelium from the mouth, &c.

6. Cuticular epithelium, angular and irregular, without apparent nuclei.

7. Striped muscular fibre.

8. Tip of a feather.

9. Barblets of ditto, more highly magnified.

10. Scales of Insects. Besides the Lepidotera—namely, the Moths and Butterflies, numerous other insects are furnished with scales. Thus they form a velvety coat on the Anthracidæ and Bombylidæ, but are more distinctly scaly on bodies of many of the Curculionidæ, Melolonthidæ, Clavicornes, Lepismidæ, Poduridæ, and on the wings of the Culicidæ (Siebold).

BACTERIA.

MICROCOCCUS.
a
b
c
d
e
BACTERIUM.
f
g
h
i
k
l
BACILLUS.
m
n
VIBRIO
o
p
1
2
SPIRILLUM.
q
1
2
r
s
SPIROCHÆTA.
t

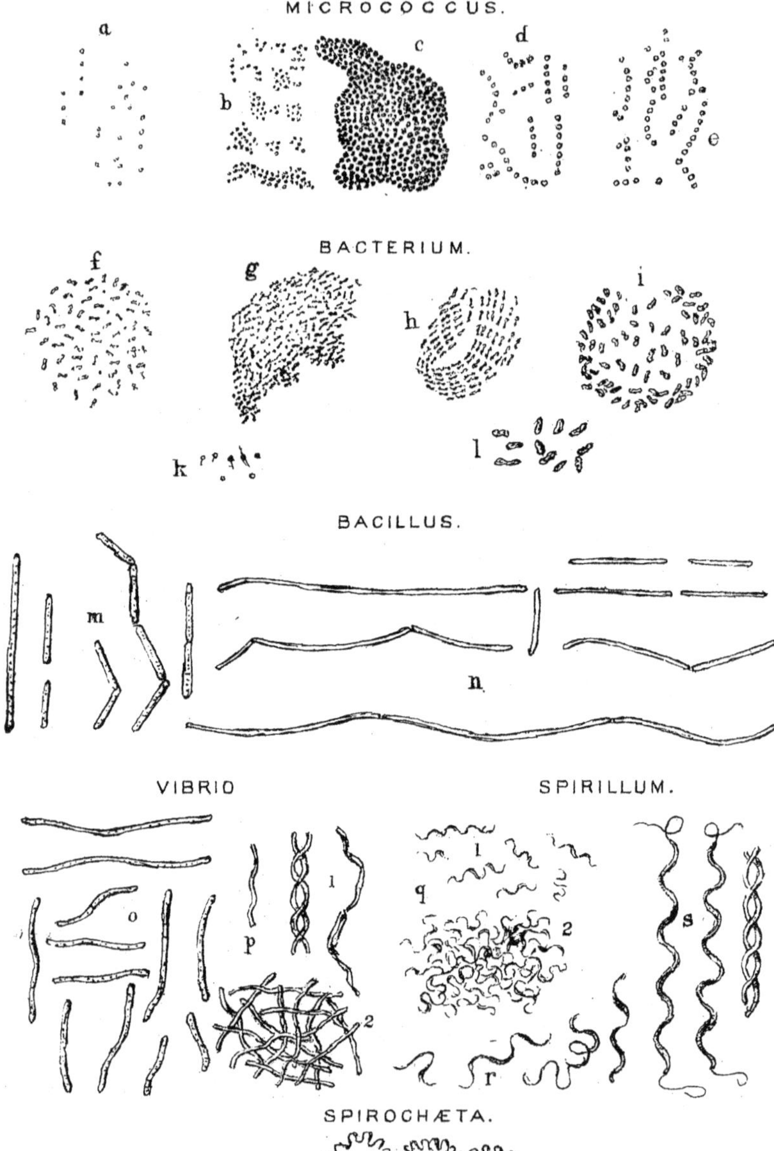

PLATE VI.

Bacteria.

MICROCOCCUS.

a. M. prodigiosus. *b*. M. vaccinæ. *c*. M. crepusculum. *d*. M. ureæ. *e*. An allied species.

BACTERIUM.

f. B. termo, free. *g*. Ditto in the zooglœa form. *h*. Ditto in linear series.

l. B. lineola, free. *i*. Ditto in the zooglœa form. *k*. Bacteria with highly refracting point.

BACILLUS.

m. B. ulnea. *n*. B. subtilis.

VIBRIO.

o. V. rugula. *p*. V. serpens; 1. free, or in twin spirals, 2, felted together.

SPIRILLUM.

q. S. tenue; 1. free, 2. felted together. *r*. S. undula. *s*. S. volutans.

SPIROCHÆTA.

t. S. plicatilis.

FRONDS WITH BACTERIA.

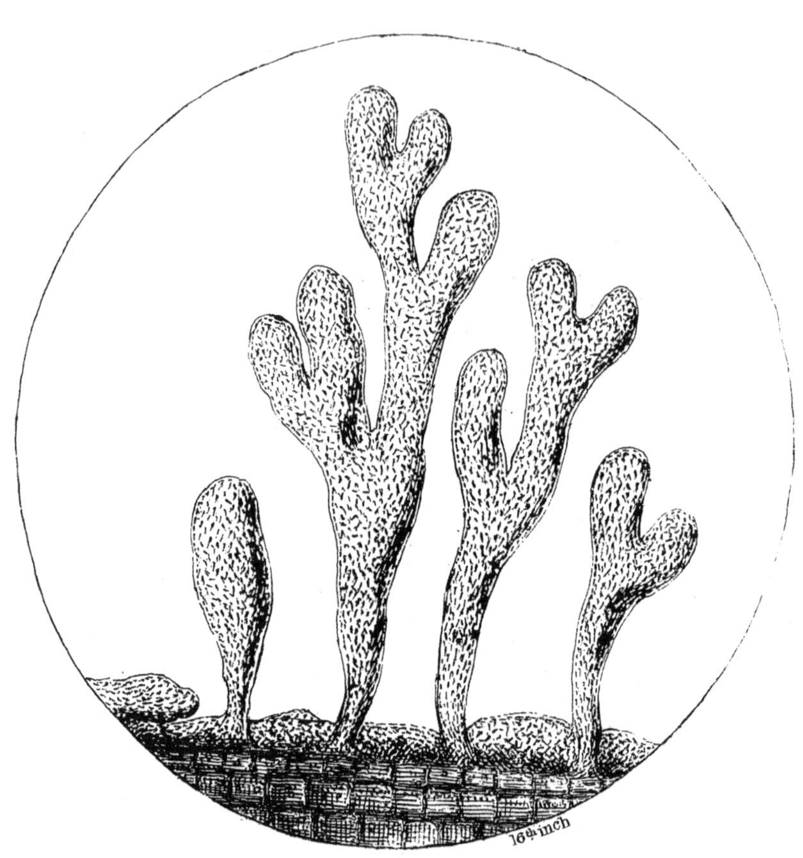

Minute Bacteroids in clavate simple or branched fronds on a spray of pond weed.

PLATE VII.

Fronds with Bacteria.

This Plate represents exceedingly minute gelatinous fronds, with embedded bacteroids growing upon a decaying portion of pond weed (Potamogeton). An encrusting layer is seen at the base from which the little fronds spring.

The great number and extreme minuteness of the molecular forms of vegetable life must still claim the attention of Hygienists, from their possible connexion with certain subtle types of disease, until our knowledge has made sufficient progress, either to accept, or reject them as efficient causes.

OSCILLATORIACEÆ.

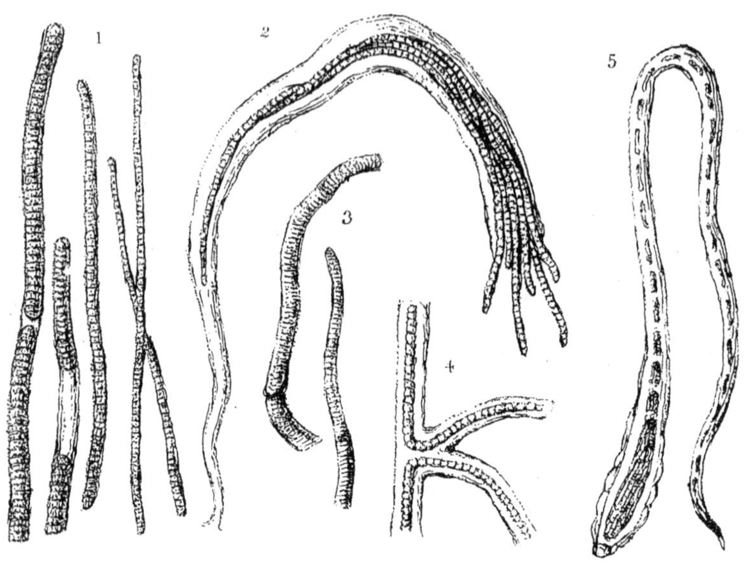

NOSTOCHACEÆ.

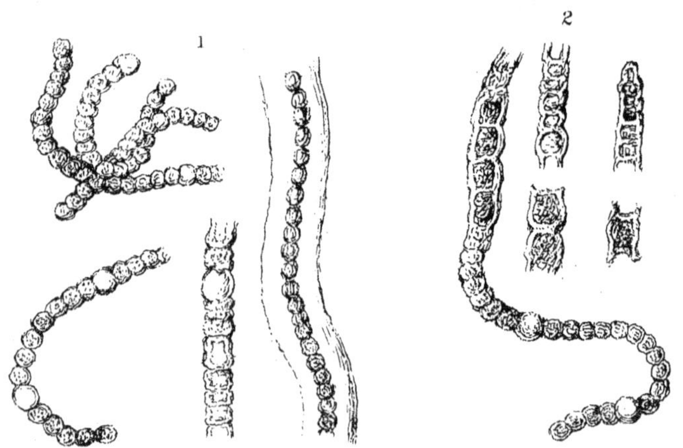

PLATE VIII.

Oscillatoriaceæ.

1. Oscillatoria autumnalis and allied species.
2. Microcoleus repens.
3. Lyngbya muralis.
4. Scytonema Myocrous.
5. Rivularia Boryana.

Nostochaceæ.

1. Nostoc commune. Several fragments showing vesicular cells to the left, and a filament in a gelatinous sheath to the right.

2. Trichormus musicola. The longer portion to the left exhibiting spermatic and vesicular cells, and the smaller segments to the right, the effect of treatment with acid.

PALMELLACEÆ.

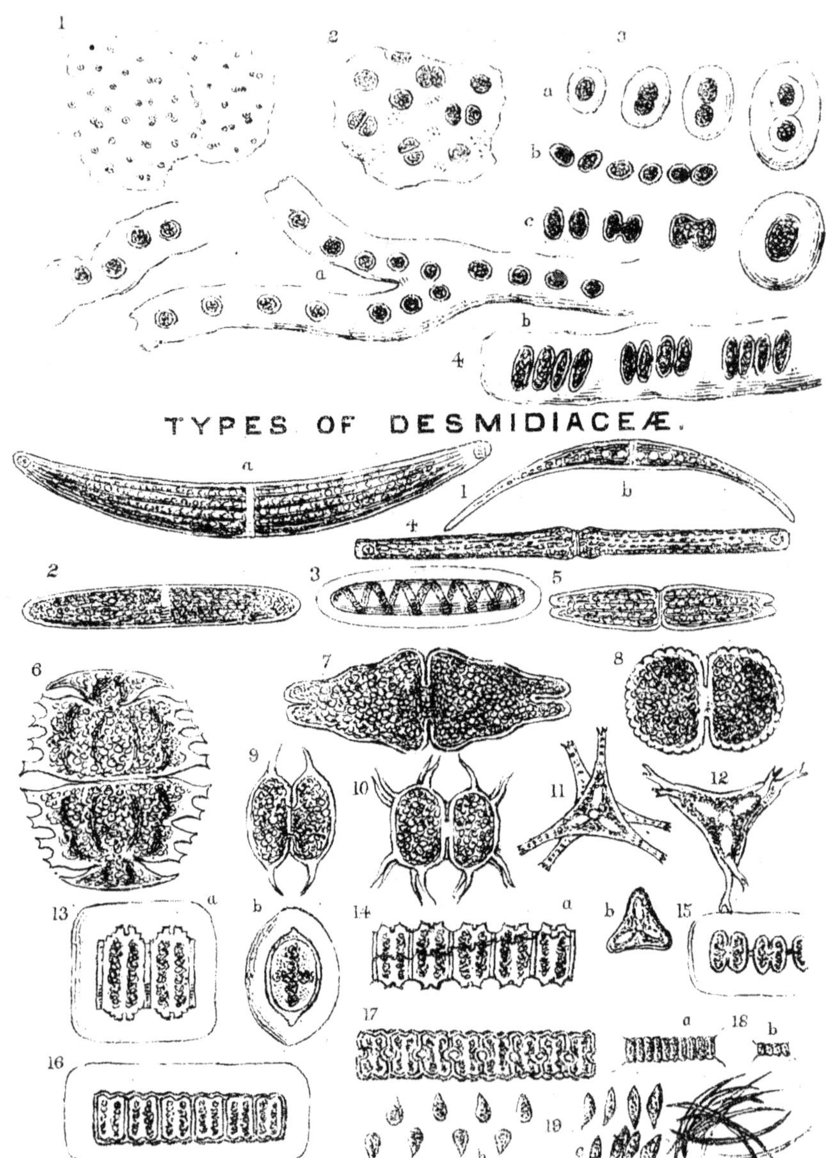

PLATE IX.

Palmellaceæ.

1. Microhaloa Ichthyoblabe. 2. Palmella cruenta.

3. Coccochloris Brebissonii. (*a*) Development and cleavage of a cell resulting in two new cells, each enclosed in a new gelatinous coat within the primary one. (*b*) Multiplication in the absence of the moisture necessary for the production of the gelatinous coat. (*c*) Approximation, union, and coalescence of two endochromes, to form a new cell, with the capability of repeating the process with a similar cell.

4. Hormospora (*a*) mutabilis, and (*b*) transversalis; which latter makes a near approach to some of the humbler Desmidiaceæ.

Types of Desmidiaceæ.

1. Closterium (*a*) lunula, (*b*) moniliformis. 2. Penium Brebissonii. 3. Spirotænia condensatum. 4. Docidium baculum. 5. Tetmemorus Brebissonii. 6. Micrasterias sp. (Fiji.) 7. Euastrum didelta. 8. Cosmarium margaritiferum. 9. Arthrodesmus convergens. 10. Xanthidium fasciculatum. 11. Staurastrum gracile. 12. Didymocladon furcigerus. 13. Didymoprium Grevillii, (*a*) front, and (*b*) side-view. 14. Desmidium Swartzii, (*a*) front, and (*b*) side-view. 15. Sphærozosma vertebratum. 16. Hyalotheca dissilens. 17. Aptogonum desmidium. 18. (*a* and *b*) Scenedesmus quadricornis. 19. (*b*) Scenedesmus obtusus, (*c*) S. obliquus. 20. Ankistrodesmus falcatus.

34 TYPES OF FRESH WATER DIATOMACEÆ.

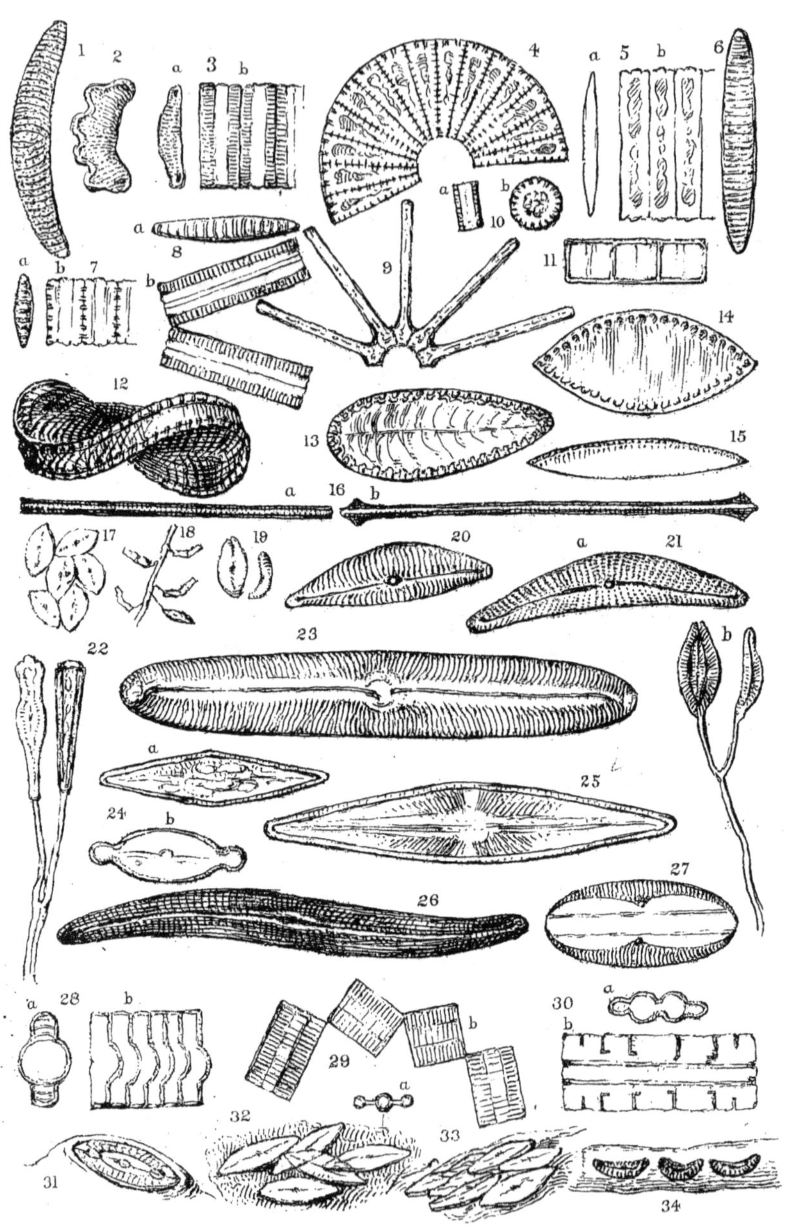

PLATE X.

Thirty-four Types of Fresh-Water Diatomaceæ.

1. Epithemia turgida. 2. Eunotia tetraodon. 3. Himantidium pectinale (*a* side, and *b* front view). 4. Meridion circulare. 5. Fragilaria capucina. 6. Denticula elegans. 7. Odontidium turgidum (*a* side, and *b* front view). 8. Diatoma vulgare (*a* side, and *b* front view). 9. Astrionella formosa. 10. Cyclotella opercula. 11. Melosira varians. 12. Campylodiscus spiralis. 13. Surirella splendida. 14. Sphynctocystis elliptica. 16. Synedra (*a* splendens, *b* capitata). 17. Cocconeis pediculus. 18. Achnanthes minutissima. 19. Achnanthidium microcephalum. 20. Cymbella Ehrenbergii. 21. Cocconema lanceolatum, *a* and *b* (*a*, single frustule highly magnified). 22. Gomphonema acuminatum. 23. Pinnularia grandis. 24. (*a*) Navicula cuspidata, (*b*) N. sphærophera. 25. Stauroneis acuta. 26. Gyrosigma attenuatum. 27. Amphora ovalis. 28. Tetracyclus lacustris (*a* side, *b* front view). 29. Tabellaria floccosa (*a* side, *b* front view). 30. Terpsinoe musica (*a* side, *b* front view). 31. Mastogloia lanceolata. 32. Frustulia saxonica. 33. Colletonema vulgare. 34. Encyonema paradoxum.

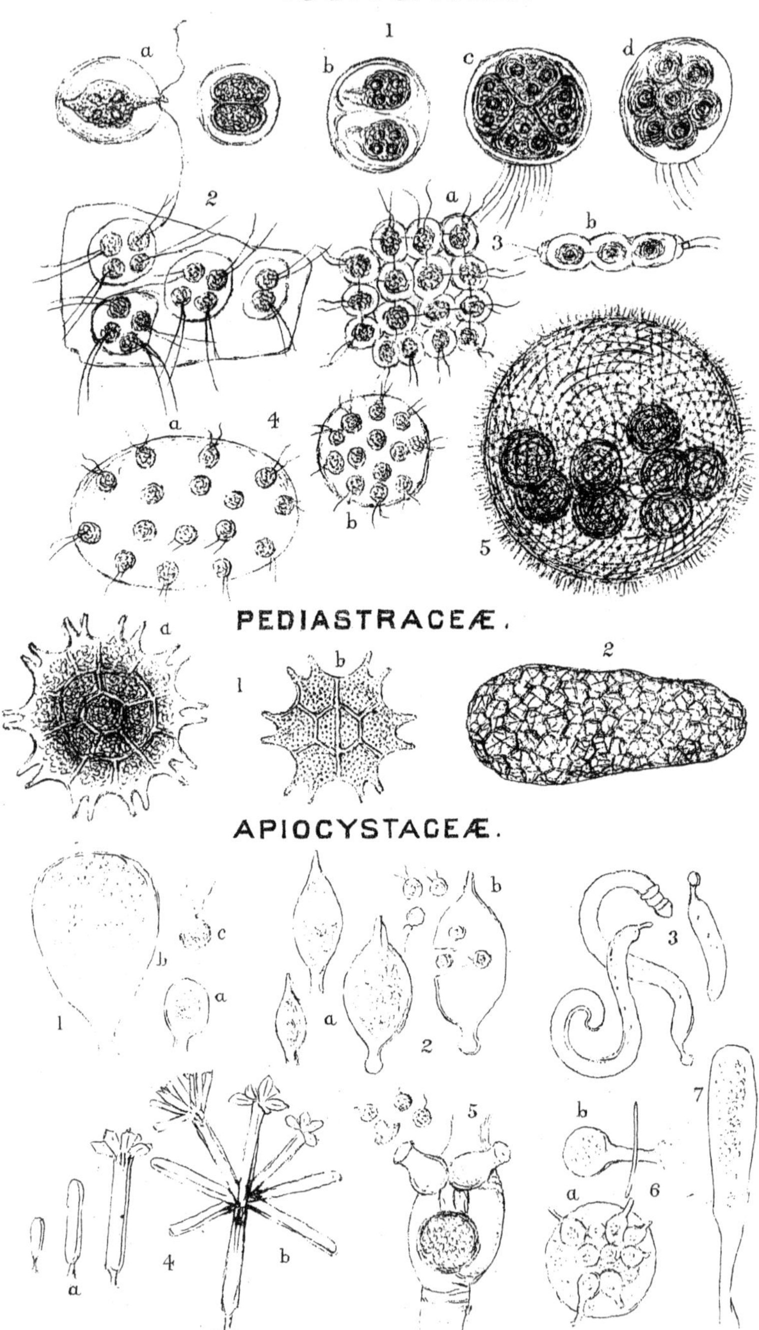
VOLVOCACEÆ.
PEDIASTRACEÆ.
APIOCYSTACEÆ.

PLATE XI.

Volvocaceæ.

1. Protococcus viridis, *a*, a single motile cell, and a stationary one undergoing cleavage of the endochrome; *b*, two resulting cells; *c*, cleavage into four, and *d*, into eight new cells, within the primary one. 2. Tetraspora gelatinosa. 3. Gonium pectorale (*a* seen in face, *b* seen edgewise). 4. Pandorina morum (*a* side view, and *b* end view). 5. Volvox globator.

Pediastraceæ (Provisional).

1. Pediastrum. *a*, Boryanum. *b*, granulatum.
2. Hydrodictyon utriculatum.

Apiocystaceæ (Provisional).

1. Apiocystis Brauniana (*a* young, *b* zoospore). 2. Hydrocytium acuminatum (*a*, stages of growth, *b*, shedding zoospores). 3. Ophiocytium majus. 4. Sciadium arbuscula (*a*, stages of development, *b*, complete form). 5. Chytridium Olla, on a filament of Œdogonium, one dehiscing and discharging monad-like zoospores. 6. Pythium entophytum (*a*, an immature cluster in a cell of Chlorosphæra, *b*, one perforating the cell-wall and discharging its contents). 7. Codiolum gregarium.

SIPHONACEÆ.

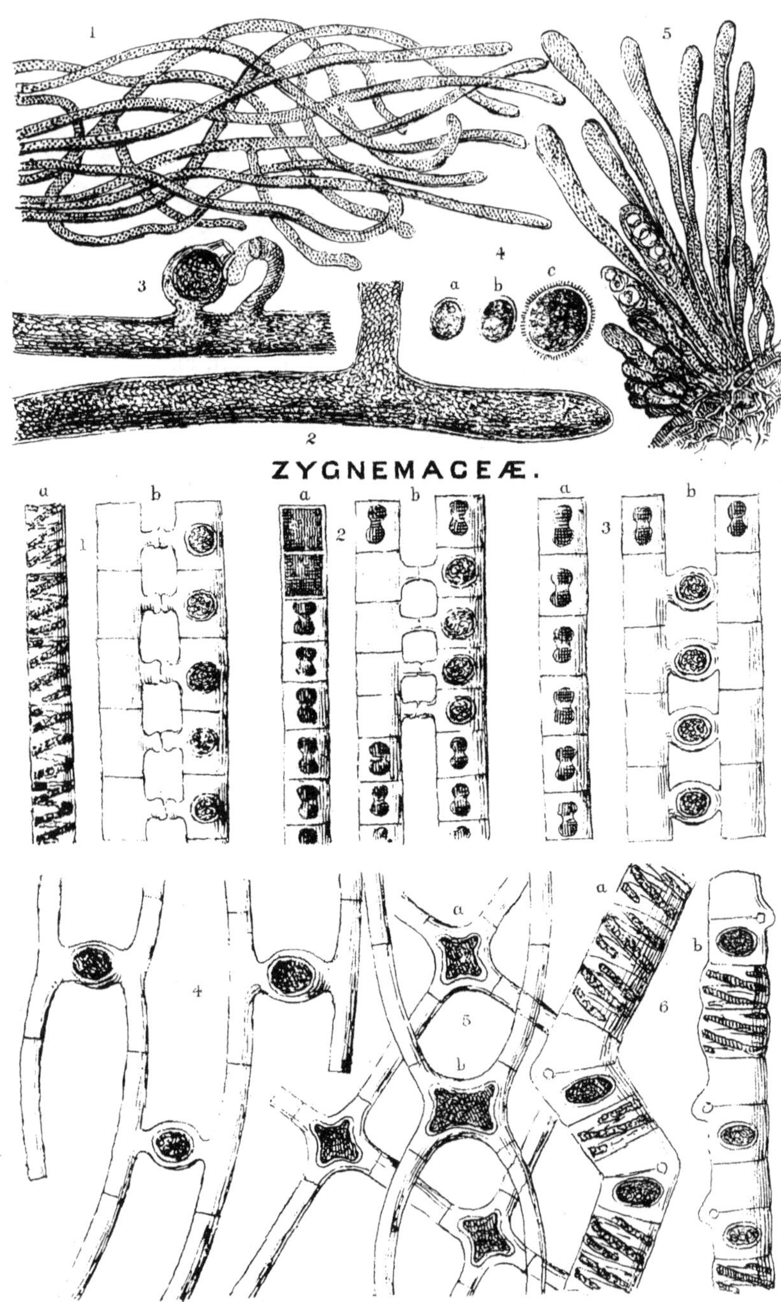

PLATE XII.

Siphonaceæ.

1. **Vaucheria** Ungeri. 2. Portion more highly magnified.

3. Sporange and antheridium. 4. *a* and *b* stages in the development of *c*, the ciliated spore of Vaucheria.

5. Achlya prolifera, with its mycelioid rootlets growing upon the dead body of a small fly.

Zygnemaceæ.

1. Spirogyra. 2. Zygnema. 3. Zygogonium.

In all three cases the simple filament is shown at *a*, and the mode of conjugation at *b*.

4. Mesocarpus. 5. Staurocarpus. 6. Rhynchonema. *a* and *b* in the two latter figures merely indicate different species.

CONFERVACEÆ (a & b) ŒDOGONIACEÆ (c & d) CHÆTOPHORACEÆ (e).

PLATE XIII.

Confervaceæ, Œdogoniaceæ, and Chætophoraceæ.

a. Conferva floccosa. *b.* Cladophora crispata. *c.* Species of Œdogonium. *d.* Bulbochæte setigera. *e.* Chætophora elegans. Amongst the Diatomaceæ introduced in this Plate may be noticed—Long prismatic Synedræ, Tabellaria floccosa, wedge-shaped and stalked Gomphonemæ, with the little bent frustules of Achnanthes minutissima. A spray of pond weed forms the theatre of this microscopic vegetation.

RHIZOPODA.

RADIOLARIA.

1. *Actinophrys.* 2. *Acanthocystis* 3. *Clathrulina.*

RETICULARIA

Pleurophrys *Amphitrema.* *Gromia.*

LOBOSA

1. *Trinema acinus* 2. *Euglypha tuberculata.* 3. *E. alveolata.*
4. *Difflugia, two forms.* 5. *Arcella Vulgaris, from above.* 6. *Cyphidium.* 7. *Amœba, several forms.*

PLATE XIV.

Rhizopoda.

RADIOLARIA.

1. Actinophrys; (*a*) Eichornii; (*b*) sol.; (*c*) ditto young. 2. Acanthocystis turfacea; (*a*) full grown, (*b*) young. 3. Clathrulina elegans.

RETICULARIA.

1. Gromia fluviatilis. 2. Pleurophrys amphitremoides. 3. Amphitrema Wrightianum.

LOBOSA.

1. Trinema acinus. 2. Euglypha tuberculata. 3. E. alveolata. 4. Difflugia (*a*) spinosa, (*b*) proteiformis. 5. Arcella vulgaris. 6. Cyphidium aureolum. 7. Amœba, (*a*) ramosa, (*b*) radiosa, (*c*) young of diffluens.

Pl. XV.

INFUSORIA.

FLAGELLATA. MONADINA.

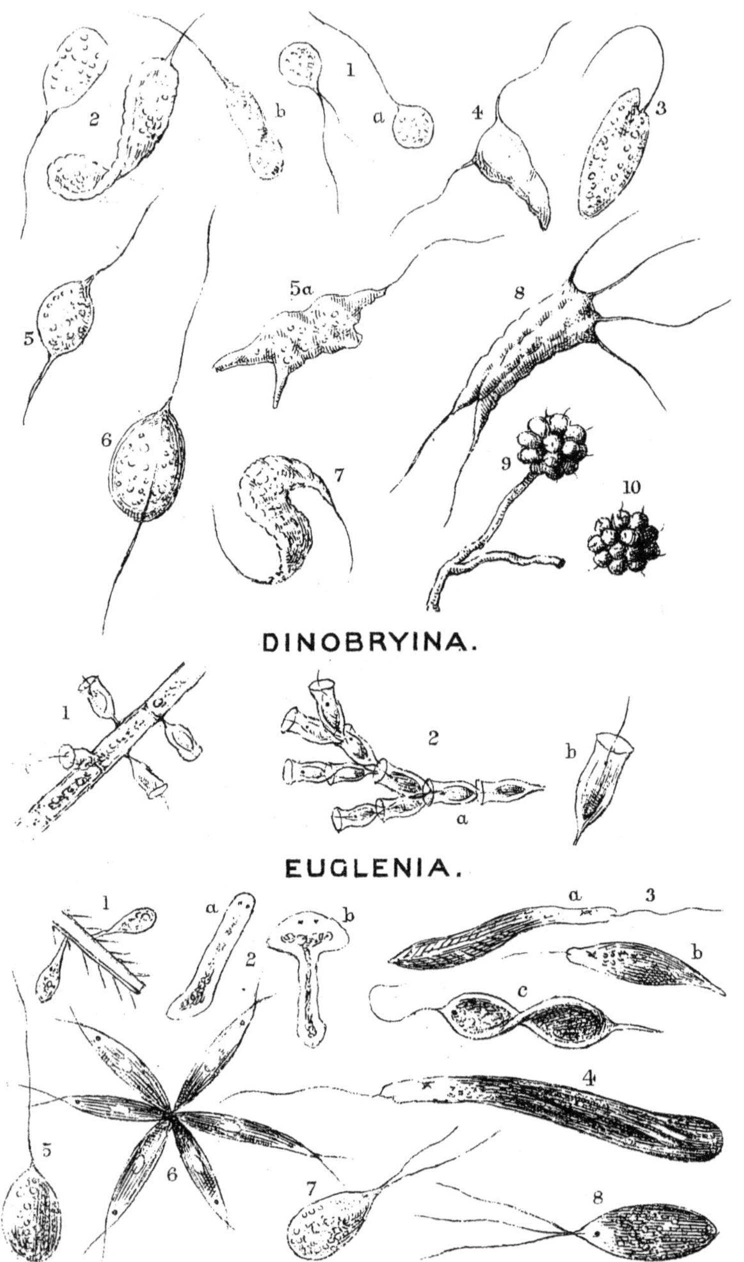

PLATE XV.

Infusoria.

FLAGELLATA. MONADINA.

1. Monas (*a*) lens, (*b*) attenuata. 2. Cyclidium (*a*) abscissum, (*b*) distortum. 3. Chilomonas granulosa. 4. Amphimonas dispar. 5. Cercomonas (*a*) longicauda, (*b*) lobata. 6. Heteromita exigua. 7. Trepomonas agilis. 8. Hexamita nodulosa. 9. Anthophysa Mülleri. 10. Uvella glaucoma.

DINOBRYINA.

1. Epipyxis utriculus. 2. Dinobryon sertularia; (*a*) normal state, (*b*) separate cell more highly magnified.

EUGLENIA.

1. Colacium vesiculosum. 2. Distigma; (*a*) proteus, (*b*) viride. 3. Euglena; (*a*) spirogyra, (*b*) viridis, (*c*) longicauda. 4. Amblyophis viridis. 5. Peranema globulosa. 6. Chlorogonium euchlorum. 7. Zygoselmis inæqualis. 8. Polyselmis viridis.

FLAGELLATA (Cont^d) THECAMONADINA.

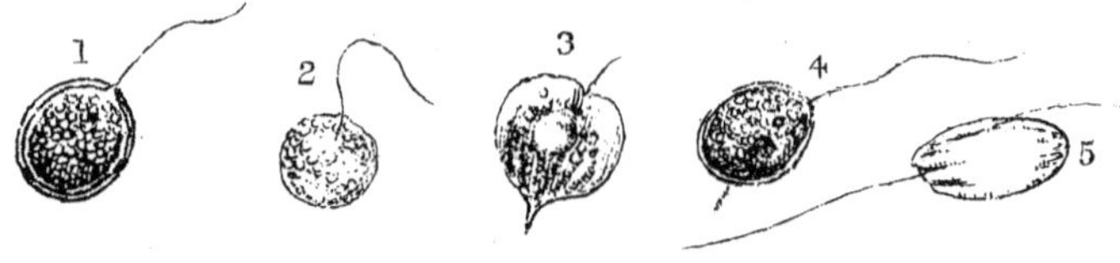

PERIDINÆA.

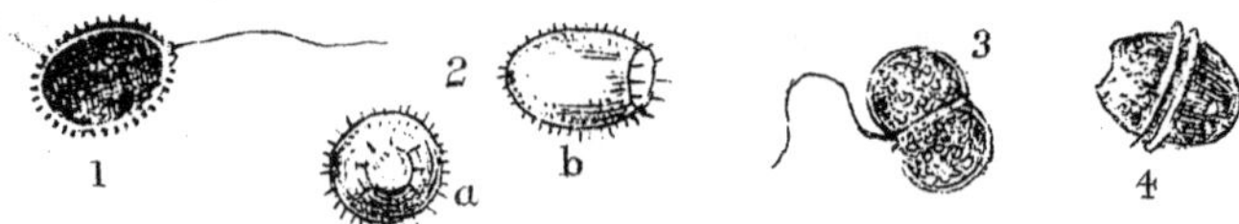

CILIATA.

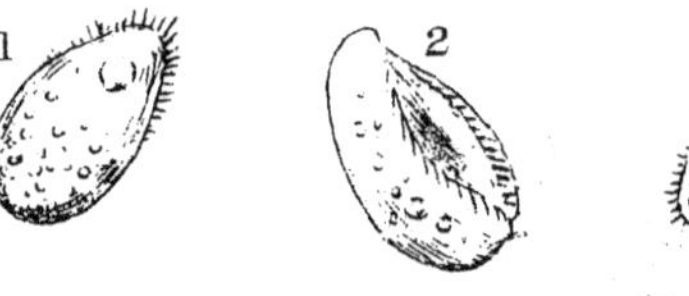

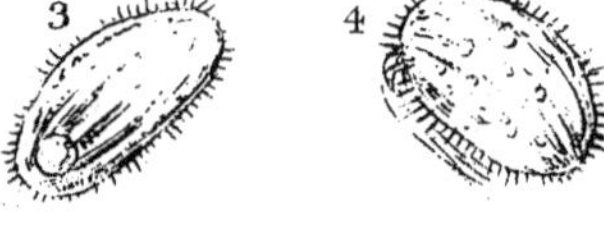

TRICHODINA.

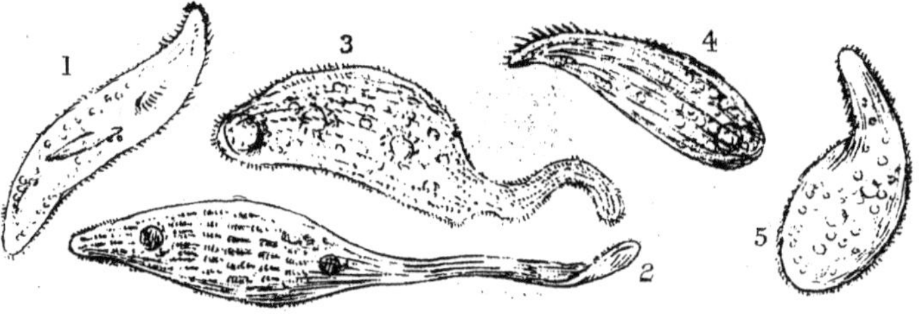

KERONIA.

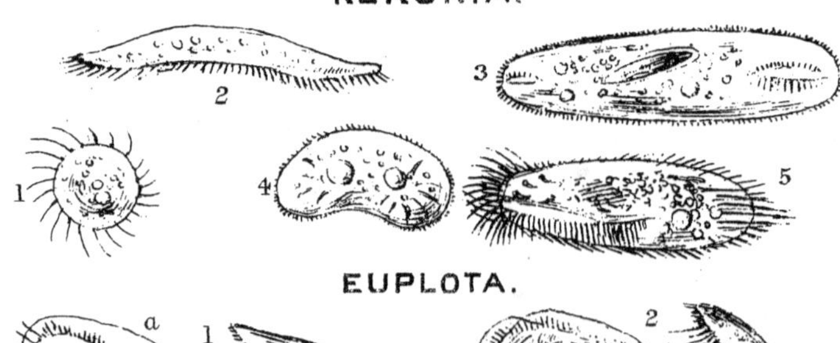

EUPLOTA.

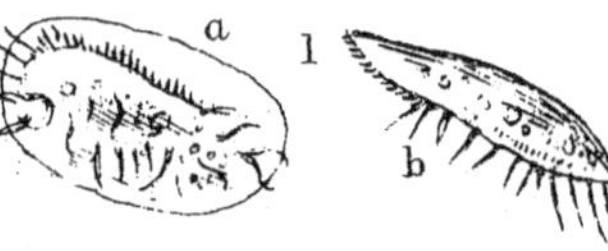

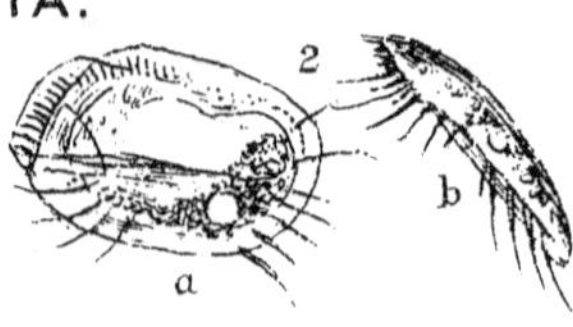

PLATE XVI.

THECAMONADINA.

1. Trachelomonas volvocina. 2. Cryptomonas globulus. 3. Phacus pleuronectes. 4. Crumenula texta. 5. Anisonema sulcata.

PERIDINÆA.

1. Chætoglena sp. 2. Chætotyphla armata; (*a*) end, and (*b*) side view. 3. Glenodinium cinctum. 4. Peridinium cinctum.

CILIATA. ENCHELIA.

1. Acomia vitrea. 2. Gastrochæta fissa. 3. Enchelys nodulosa. 4. Alyscum saltans.

TRICHODINA.

1. Pelecida rostrum. 2. Dileptus folium. 3. Trachelius anas. 4. Acineria incurvata. 5. Trichoda angulata.

KERONIA.

1. Halteria grandinella. 2. Oxytricha gibba. 3. Urostyla grandis. 4. Kerona polyporum. 5. Stylonychia histrio (lanceolata?).

EUPTOTA.

1. Himantophorus charon; (*a*) front, and (*b*) side view. 2. Euplotes vannus, (*a*) front, and (*b*) side view.

Pl. XVI.

CILIATA (Contd) PARAMECIA.

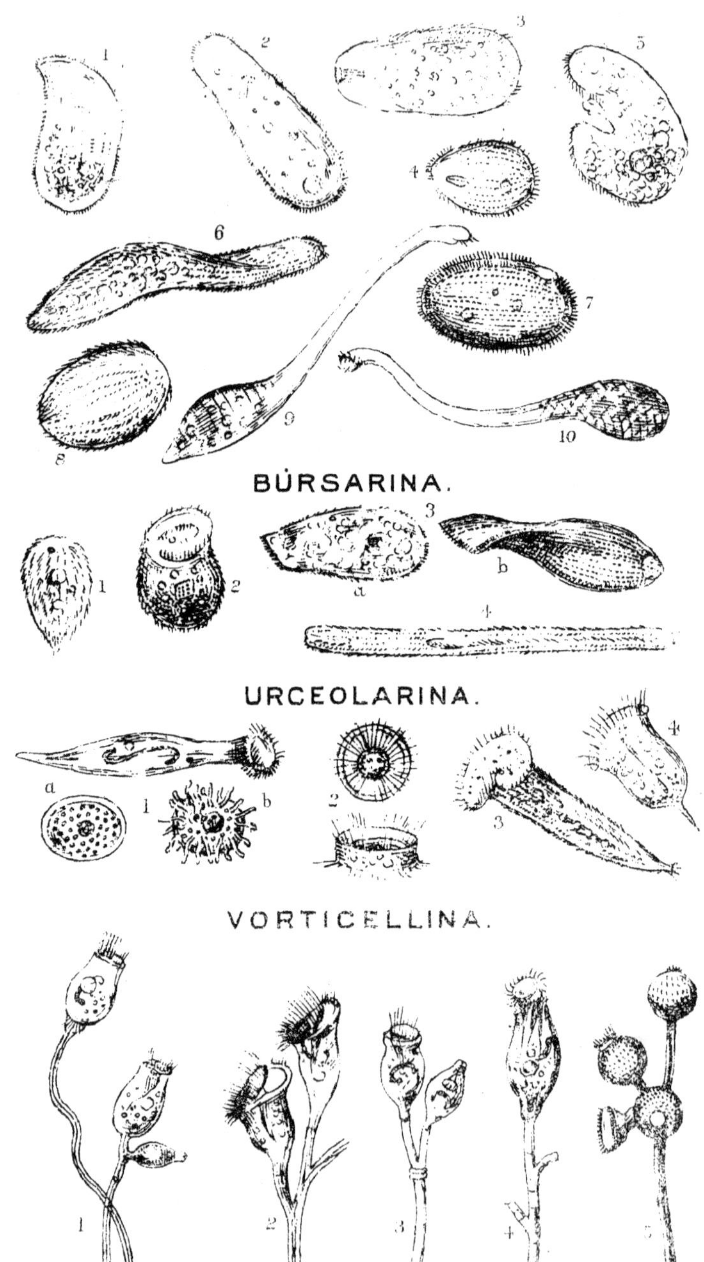

PLATE XVII.

PARAMECIA.

1. Chilodon cucullulus. 2. Nassula elegans. 3. Prorodon teres. 4. Glaucoma scintillans. 5. Colpoda cucullus. 6. Paramecium aurelia (three-quarter-view). 7. Panophrys crysalis. 8. Holophrya ovum. 9. Trachelocera olor. 10. Lacrymaria proteus.

BURSARINA.

1. Ophryoglena acuminata. 2. Bursaria vorticella. 3. Leucophrys (*a*) patula, (*b*) spathula, Ehr. (Spathidium hyalinum) Du. 4. Spirostomum ambiguum.

URCEOLARINA.

1. Ophrydium versatile, showing an animal in the extended state, and (*a*) encysted, (*b*) the supposed Acineta form. 2. Urceolaria pediculus (Trichodina). 3. Stentor cœruleus, with internal germs. Urocentrum turbo.

VORTICELLINA.

1. Vorticella microstoma. 2. Carchesium polypinum. 3. Epistylis crassicollis. 4. Opercularia articula. 5. Zoothamnium arbuscula.

CILIATA (*Cont^d*) SYMMETRICAL FORMS.

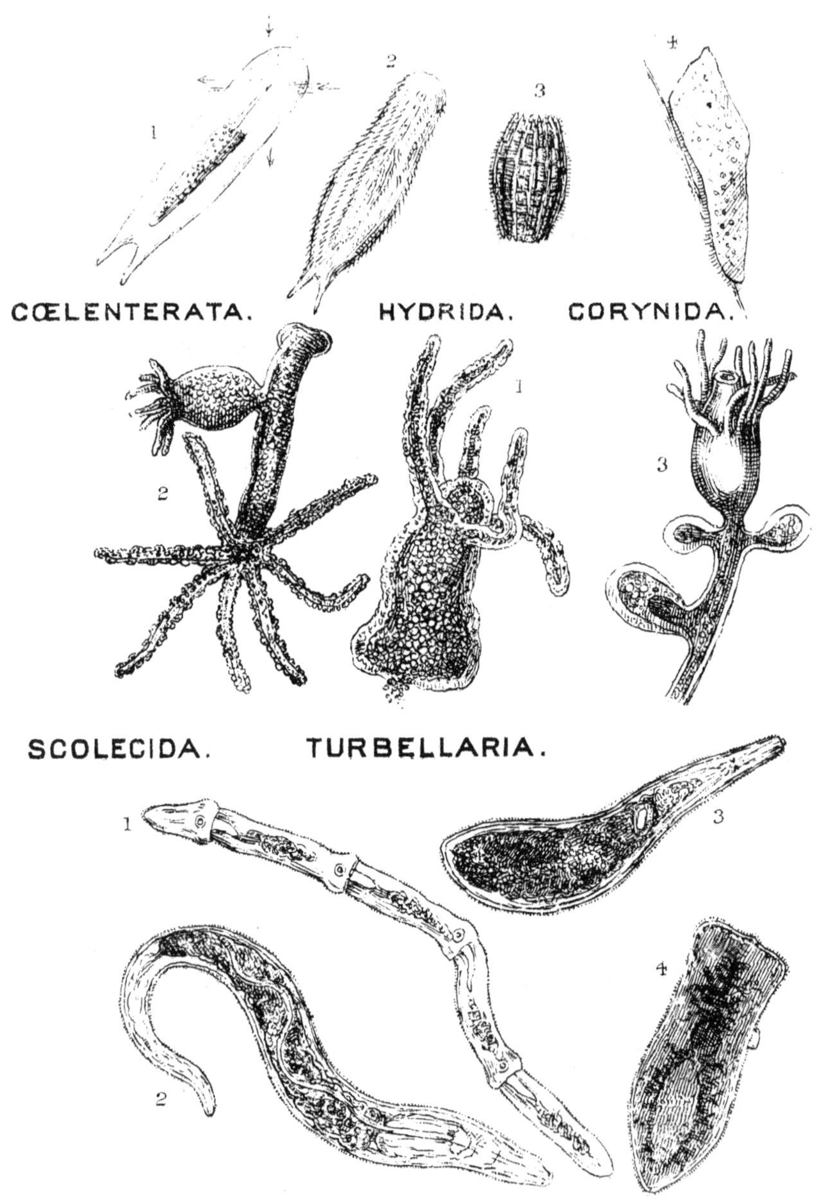

PLATE XVIII.

CILIATA—*continued.*

Symmetrical Forms.

1. Ichthydium Podura. 2. Chætonotus Larus. 3. Coleps hirtus. 4. Planariola rubra.

Cœlenterata.

1. Hydra viridis. 2. H. vulgaris. 3. Cordylophora lacustris.

Turbellaria.

1. Derostomum. 2. Prostomum. 3. Mesostomum. 4. Planaria.

SCOLECIDA (Contd) NEMATODA.

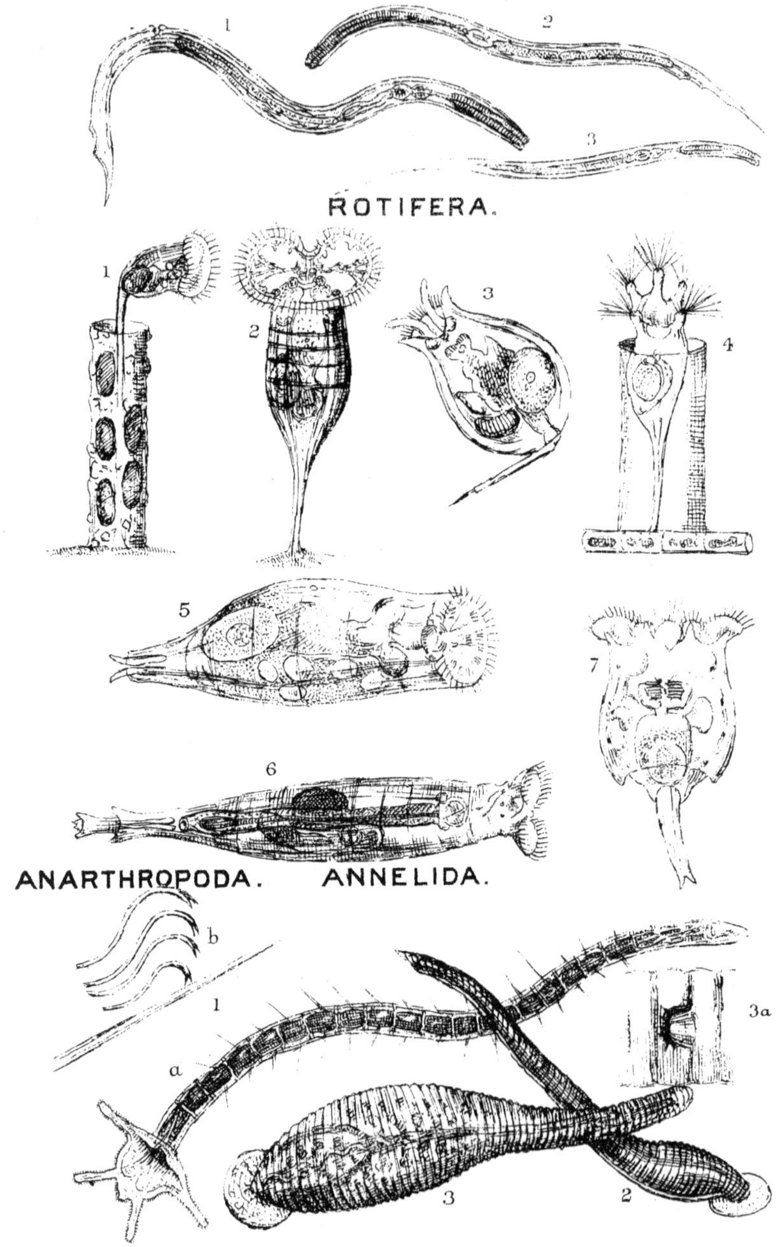

PLATE XIX.

Nematoda.

1. Anguillula (from bilge water). 2. A. aceti. 3. A. fluviatilis.

Rotifera.

1. Œcistes crystallinus. 2. Megalotrocha flavicans. 3. Monostyla quadridentata. 4. Floscularia ornata. 5. Hydatina senta. 6. Rotifer vulgaris. 7. Brachionus amphiceros.

Annelida.

1. Naid; (*a*) conformable with the Proto of Oken; (*b*) setæ, and ventral hooklets. 2. Nephelis, sp. 3. Glossiphonia bioculata; 3*a*. a dorsal chitinous tooth-like process directed backwards from the eleventh segment, over a little pit in the twelfth.

ENTOMOSTRACA.

OSTRACODA.

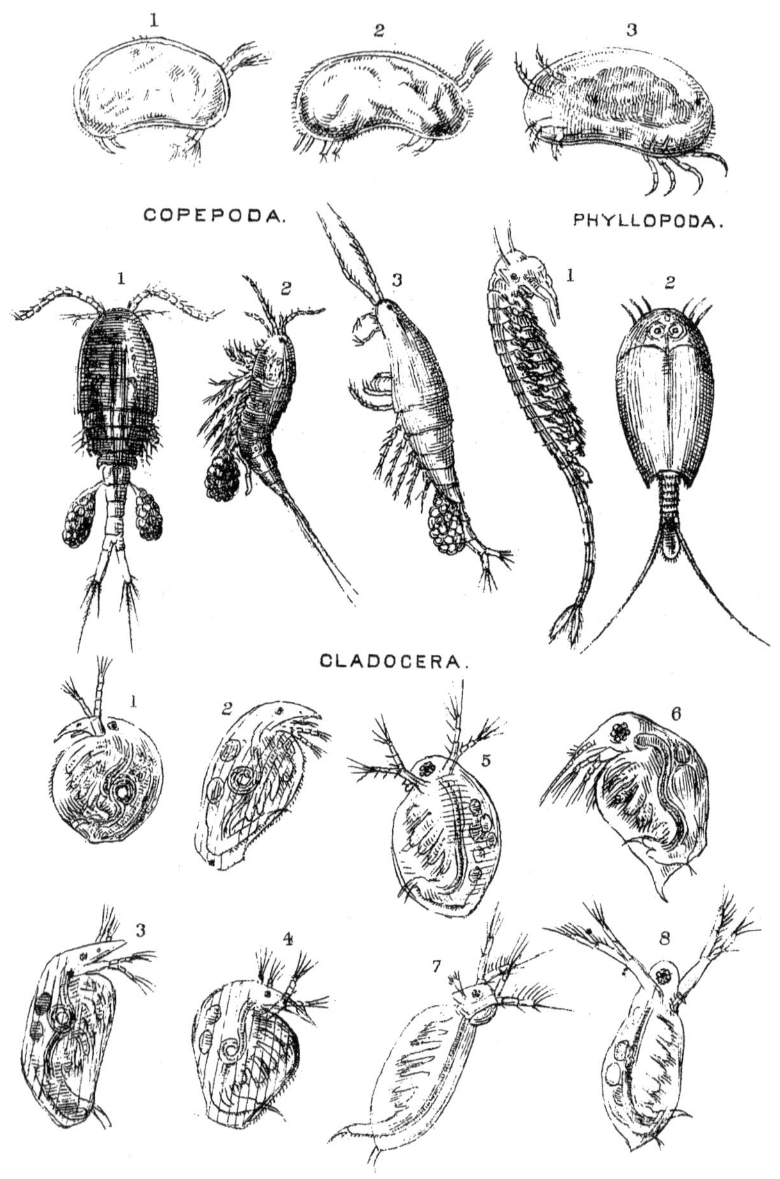

PLATE XX.

Entomostraca.

OSTRACODA.

1. Cypris tristriata. 2. Candona reptans. 3. Cythere inopinator.

COPEPODA.

1. Cyclops quadricornis. 2. Canthocamptus minutus. 3. Diaptomus castor.

PHYLLOPODA.

1. Branchipus stagnalis. 2. Lepidurus, *Leach*=Monoculus Apus of *Linnæus.*

CLADOCERA.

Lynceidæ.

1. Chydorus sphæricus. 2. Camptocercus macrourus.
3. Alona quadrangularis. 4. Pleuroxus trigonellus.

Daphnidæ.

5. Daphnia pulex. 6. Bosmina longirostris.
7. Sida crystallina. 8. Daphnella Wingii.

MALACOSTRACA. Pl. XXI.

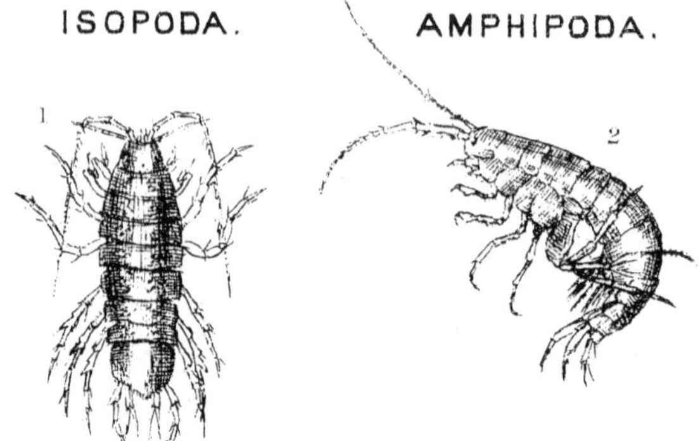

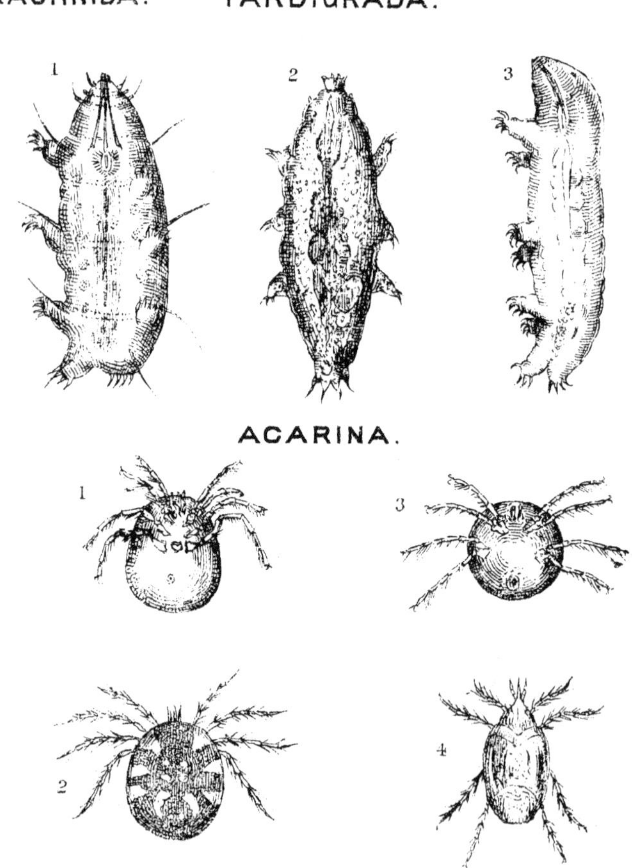

PLATE XXI.

Malacostraca.

Isopoda. 1. Asellus aquaticus.
Amphipoda. 2. Gammarus pulex.

Arachnida.

TARDIGRADA.

1. Emydium testudo. 2. Milnesium tardigrada.
3. Macrobiotus Hufelandii.

ACARINA.

1. Hydrachna globula. 2. H. geographica.

3. A more globular form in which, quite exceptionally, six eyes are present.

4. Limnochares holocericus, a crawling water mite.

INSECTA.

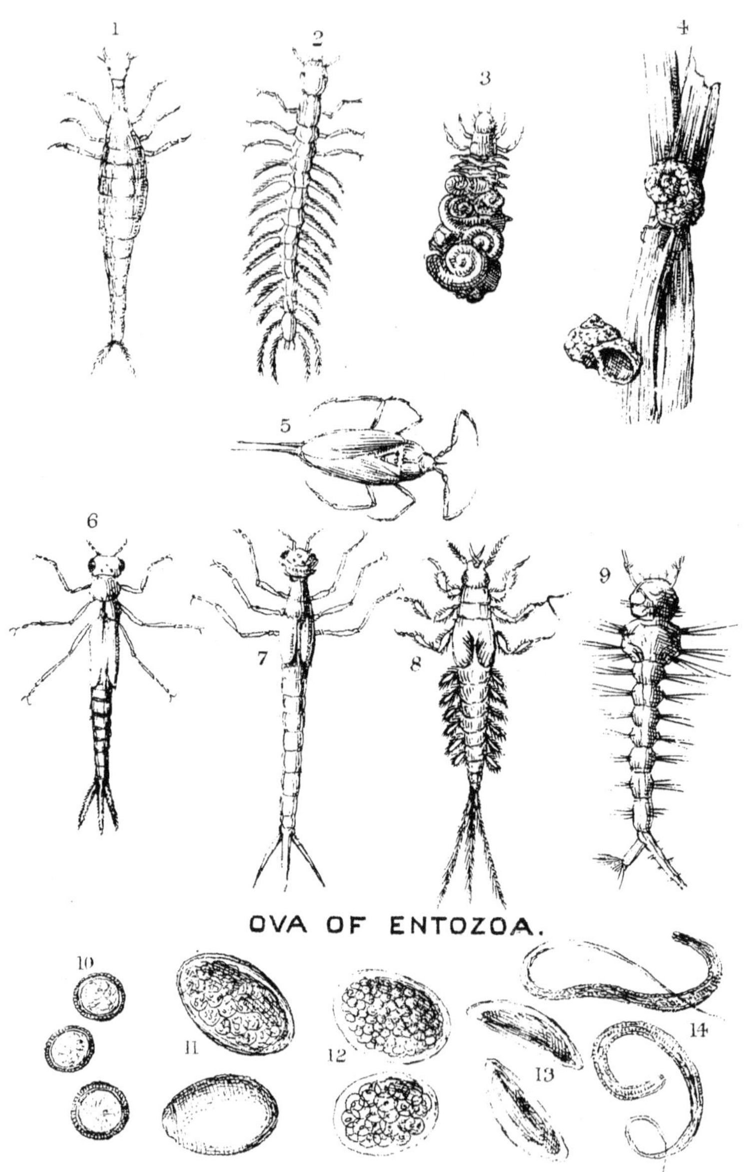

PLATE XXII.

Insecta.

COLEOPTERA.

1. Larva of Acilius sulcatus.
2. Larva of Gyrinus natator.

TRICHOPTERA.

3. Phryanea grandis in its composite case.
4. The form named *Thelidomus* by Mr. Swainson, who mistook the little built-up case for a genuine shell, and gave it a place among the Helices (snails), arranged in accordance with the "quinary system."

The case figured, from the Isle of Pines, S. W. Pacific, was made of granules of ironstone, but in some of the streams of New Caledonia, the retreat of probably the same species, is constructed of little amethysts.

HÆMIPTERA.

5. Pupa of Nepa (water scorpion).

NEUROPTERA.

6. Pupa of Agrion puella.
7. Pupa of Calepteryx virgo.
8. Pupa of Ephemera vulgata.

DIPTERA.

9. Larva of the Gnat Culex pipiens.

OVA OF ENTOZOA.

10. Of Tænia mediocanellata.
11. Of Fasciola hepatica.
12. Of Ascaris dentata.
13. Of Bilharzia hæmatobia.
14. Young of Filaria medinensis.

WELL-WATER (NETLEY).

a. *Gelatinous fronds with Bacteria.*
b. *Monadina.*
c. *Diatomaceæ.*
d. *Star shaped cells*
e. *Ankistrodesmus.*
f. *Desmidiaceæ.*
g. *Thecomonadina.*
h. *Palmella, (minute.)*
i. *Euglena viridis.*
k. *Spore of Septoria, (fungus).*
l. *Oscillatoria.*

PLATE XXIII.

Well-Water (*Netley*).

The suspended matters represented in this Plate were obtained by setting aside a tall glass litre measure full of the water, with a disc of glass attached to a long wire at the bottom. During the first twelve hours a deposit of grosser particles was formed, with a delicate coating here and there of the gelatinous matter and bacteroids shown at *a*. In twelve hours more this coating had become more consistent, and at the end of forty-eight hours was so firmly adherent as to require some force to remove it, with the mineral particles, resting-spores of algæ, and organic débris of different kinds embedded in it.

In the little bays and creeks of this gelatinous substance the loosened and detached Bacteria were in active motion, and interspersed with Monads (*b*) of minute size.

Navicula, Synedra, and other Diatoms (*c*), were free in the field, or often projecting from the amorphous débris. The little green star-like bodies (*d*) probably allied to the Tetrapedia, have also been noticed in other specimens obtained from a deep source, and are evidently identical with those figured in Plate 4, illustrating the Reports made to the Directors of the London (Watford) Spring-Water Company by Drs. Lankester and Redfern.

The remaining objects are sufficiently explained in the references attached to the Plate.

BOG-WATER.

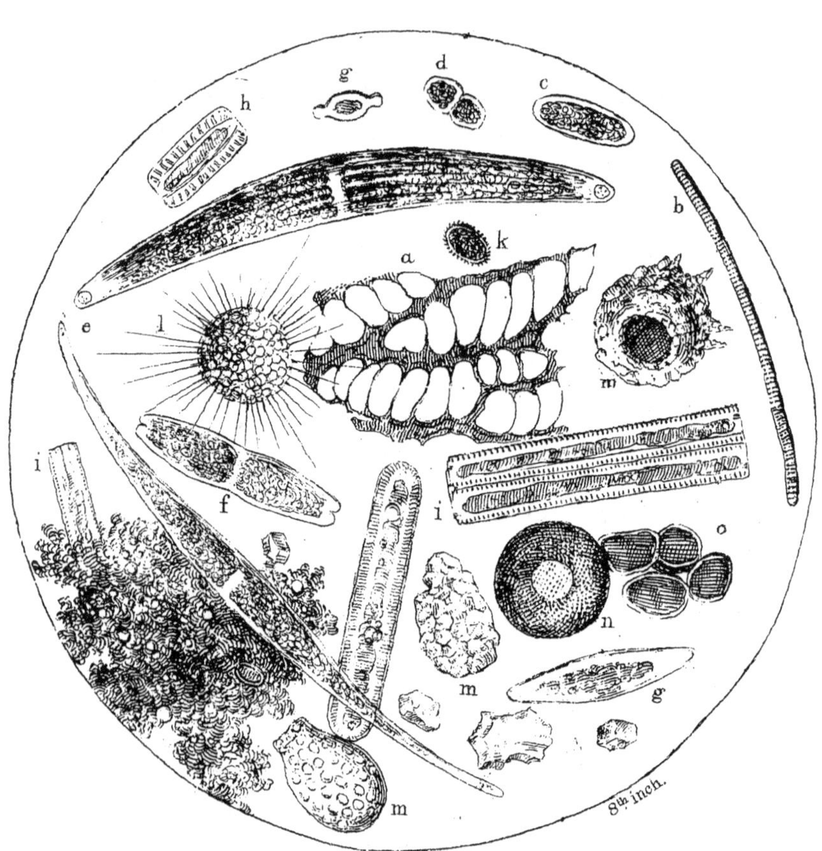

a. *Particle of bog moss.*
b. *Oscillatoria.*
c. *Penium.*
d. *Cosmarium.*
e. *Closterium.*
f. *Tetmemorus.*
g. *Navicula.*
h. *Surirella.*
i. *Pinnularia.*
k. *Chætoglena.*
l. *Actinophrys.*
m. *Difflugia.*
n. *Arcella.*
o. *Brown Vegetable cells.*

PLATE XXIV.

Bog Water.

The specimen of water here represented was taken from the swampy ground near Miller's Pond, Sholing, Southampton. It was very rich in Rhizopoda, Infusoria, Oscillatorians, and Desmids large and small, and the beautiful Pinnularia grandis, which is so plentiful in all the surrounding district, but chiefly in stagnant and impure water.

INDEX.

THE END.

LONDON:
SAVILL, EDWARDS AND CO., PRINTERS, CHANDOS STREET,
COVENT GARDEN.

www.ingramcontent.com/pod-product-compliance
Lightning Source LLC
LaVergne TN
LVHW021422110826
845150LV00007B/2049

* 9 7 8 1 4 2 5 5 1 4 3 0 3 *